# Convolutional Neural Networks in Visual Computing

# DATA-ENABLED ENGINEERING

SERIES EDITOR

**Nong Ye**

Arizona State University, Phoenix, USA

PUBLISHED TITLES

**Convolutional Neural Networks in Visual Computing: A Concise Guide**
Ragav Venkatesan and Baoxin Li

# Convolutional Neural Networks in Visual Computing
## A Concise Guide

By
Ragav Venkatesan and Baoxin Li

CRC Press
Taylor & Francis Group
Boca Raton  London  New York

CRC Press is an imprint of the
Taylor & Francis Group, an **informa** business

CRC Press
Taylor & Francis Group
6000 Broken Sound Parkway NW, Suite 300
Boca Raton, FL 33487-2742

© 2018 by Taylor & Francis Group, LLC
CRC Press is an imprint of Taylor & Francis Group, an Informa business

No claim to original U.S. Government works

Printed on acid-free paper

International Standard Book Number-13: 978-1-4987-7039-2 (Hardback); 978-1-138-74795-1 (Paperback)

**Library of Congress Cataloging-in-Publication Data**
Names: Venkatesan, Ragav, author. | Li, Baoxin, author.
Title: Convolutional neural networks in visual computing : a concise guide /
Ragav Venkatesan, Baoxin Li.
Description: Boca Raton ; London : Taylor & Francis, CRC Press, 2017. |
Includes bibliographical references and index.
Identifiers: LCCN 2017029154| ISBN 9781498770392 (hardback : alk. paper) |
ISBN 9781315154282 (ebook)
Subjects: LCSH: Computer vision. | Neural networks (Computer science)
Classification: LCC TA1634 .V37 2017 | DDC 006.3/2--dc23
LC record available at https://lccn.loc.gov/2017029154

**Visit the Taylor & Francis Web site at**
http://www.taylorandfrancis.com

**and the CRC Press Web site at**
http://www.crcpress.com

To Jaikrishna Mohan, for growing up with me;
you are a fierce friend, and my brother.
and to Prof. Ravi Naganathan for helping me grow up;
my better angels have always been your philosophy and principles.

—**Ragav Venkatesan**

To my wife, Julie,
for all your unwavering support over the years.

—**Baoxin Li**

# Contents

# Preface

Deep learning architectures have attained incredible popularity in recent years due to their phenomenal success in, among other applications, computer vision tasks. Particularly, convolutional neural networks (CNNs) have been a significant force contributing to state-of-the-art results. The jargon surrounding deep learning and CNNs can often lead to the opinion that it is too labyrinthine for a beginner to study and master. Having this in mind, this book covers the fundamentals of deep learning for computer vision, designing and deploying CNNs, and deep computer vision architecture. This concise book was intended to serve as a beginner's guide for engineers, undergraduate seniors, and graduate students who seek a quick start on learning and/or building deep learning systems of their own. Written in an easy-to-read, mathematically nonabstruse tone, this book aims to provide a gentle introduction to deep learning for computer vision, while still covering the basics in ample depth.

The core of this book is divided into five chapters. Chapter 1 provides a succinct introduction to image representations and some computer vision models that are contemporarily referred to as *hand-carved*. The chapter provides the reader with a fundamental understanding of image representations and an introduction to some linear and non-linear feature extractors or representations and to properties of these representations. Onwards, this chapter also demonstrates detection of some basic image entities such as edges. It also covers some basic machine learning tasks that can be performed using these representations. The chapter concludes with a study of two popular non-neural computer vision modeling techniques.

Chapter 2 introduces the concepts of regression, learning machines, and optimization. This chapter begins with an introduction to supervised learning. The first learning machine introduced is the linear regressor. The first solution covered is the analytical solution for least squares. This analytical solution is studied alongside its maximum-likelihood interpretation. The chapter moves on to nonlinear models through basis function expansion. The problem of overfitting and generalization through cross-validation and regularization is further introduced. The latter part of the chapter introduces optimization through gradient descent for both convex and nonconvex error surfaces. Further expanding our study with various types of gradient descent methods and the study of geometries of various regularizers, some modifications to the basic gradient descent method, including second-order loss minimization techniques and learning with momentum, are also presented.

Chapters 3 and 4 are the crux of this book. Chapter 3 builds on Chapter 2 by providing an introduction to the Rosenblatt perceptron and the perceptron learning algorithm. The chapter then introduces a logistic neuron and its activation. The single neuron model is studied in both a two-class and a multiclass setting. The advantages and drawbacks of this neuron are studied, and the XOR problem is introduced. The idea of a multilayer neural network is proposed as a solution to the XOR problem, and the backpropagation algorithm, introduced along with several improvements, provides some pragmatic tips that help in engineering a better, more stable implementation. Chapter 4 introduces the *convpool* layer and the CNN. It studies various properties of this layer and analyzes the features that are extracted for a typical digit recognition dataset. This chapter also introduces four of the most popular contemporary CNNs, AlexNet, VGG, GoogLeNet, and ResNet, and compares their architecture and philosophy.

Chapter 5 further expands and enriches the discussion of deep architectures by studying some modern, novel, and pragmatic uses of CNNs. The chapter is broadly divided into two contiguous sections. The first part deals with the nifty philosophy of using downloadable, pretrained, and off-the-shelf networks. Pretrained networks are essentially trained on a wholesome dataset and made available for the public-at-large to *fine-tune* for a novel task. These are studied under the scope of generality and transferability. Chapter 5 also studies the compression of these networks and alternative methods of learning a new task given a pretrained network in the form of mentee networks. The second part of the chapter deals with the idea of CNNs that are not used in supervised learning but as generative networks. The section briefly studies autoencoders and the newest novelty in deep computer vision: generative adversarial networks (GANs).

The book comes with a website (convolution.network) which is a supplement and contains code and implementations, color illustrations of some figures, errata and additional materials. This book also led to a graduate level course that was taught in the Spring of 2017 at Arizona State University, lectures and materials for which are also available at the book website.

Figure 1 in Chapter 1 of the book is an original image (original.jpg), that I shot and for which I hold the rights. It is a picture of the monument valley, which as far as imagery goes is representative of the southwest, where ASU is. The art in memory.png was painted in the style of Salvador Dali, particularly of his painting "the persistence of memory" which deals in abstract about the concept of the mind hallucinating and picturing and processing objects in shapeless forms, much like what some representations of the neural networks we study in the book are.

The art in memory.png is not painted by a human but by a neural network similar to the ones we discuss in the book. Ergo the connection to the book. Below is the citation reference.

```
@article{DBLP:journals/corr/GatysEB15a,
  author    = {Leon A. Gatys and
               Alexander S. Ecker and
               Matthias Bethge},
  title     = {A Neural Algorithm of Artistic Style},
  journal   = {CoRR},
  volume    = {abs/1508.06576},
  year      = {2015},
  url       = {http://arxiv.org/abs/1508.06576},
  timestamp = {Wed, 07 Jun 2017 14:41:58 +0200},
  biburl    = {http://dblp.unitrier.de/rec/bib/
               journals/corr/GatysEB15a},
  bibsource = {dblp computer science bibliography,
               http://dblp.org}
}
```

This book is also accompanied by a CNN toolbox based on Python and Theano, which was developed by the authors, and a webpage containing color figures, errata, and other accompaniments. The toolbox, named *yann* for "Yet Another Neural Network" toolbox, is available under MIT License at the URL http://www.yann.network. Having in mind the intention of making the material in this book easily accessible for a beginner to build upon, the authors have developed a set of tutorials using yann. The tutorial and the toolbox cover the different architectures and machines discussed in this book with examples and sample code and application programming interface (API) documentation. The *yann* toolbox is under active development at the time of writing this book, and its customer support is provided through GitHub. The book's webpage is hosted at http://guide2cnn.com. While most figures in this book were created as grayscale illustrations, there are some figures that were originally created in color and converted to grayscale during production. The color versions of these figures as well as additional notes, information on related courses, and FAQs are also found on the website.

This toolbox and this book are also intended to be reading material for a semester-long graduate-level course on Deep Learning for Visual Computing offered by the authors at Arizona State University. The course, including recorded lectures, course materials and homework assignments, are available for the public at large at http://www .course.convolution.network. The authors are available via e-mail for both queries regarding the material and supporting code, and for humbly accepting any criticisms or comments on the content of the book. The authors also gladly encourage requests for reproduction of figures, results, and materials described in this book, as long as they conform to the copyright policies of the publisher. The authors hope that readers enjoy this concise guide to convolutional neural networks for computer vision and that a beginner will be able to quickly build his/her own learning machines with the help of this book and its toolbox. We encourage readers to use the knowledge they may gain from this material for the good of humanity while sincerely discouraging them from building "Skynet" or any other apocalyptic artificial intelligence machines.

# Acknowledgments

It is a pleasure to acknowledge many colleagues who have made this time-consuming book project possible and enjoyable. Many current and past members of the Visual Representation and Processing Group and the Center for Cognitive and Ubiquitous Computing at Arizona State University have worked on various aspects of deep learning and its applications in visual computing. Their efforts have supplied ingredients for insightful discussion related to the writing of this book, and thus are greatly appreciated. Particularly, we would like to thank Parag Sridhar Chandakkar for providing comments on Chapters 4 and 5, as well as Yuzhen Ding, Yikang Li, Vijetha Gattupalli, and Hemanth Venkateswara for always being available for discussions.

This work stemmed from efforts in several projects sponsored by the National Science Foundation, the Office of Naval Research, the Army Research Office, and Nokia, whose support is greatly appreciated, although any views/conclusions in this book are solely of the authors and do not necessarily reflect those of the sponsors. We also gratefully acknowledge the support of NVIDIA Corporation with the donation of the Tesla K40 GPU, which has been used in our research.

We are grateful to CRC Press, Taylor and Francis Publications, and in particular, to Cindy Carelli, Executive Editor, and Renee Nakash, for their patience and incredible support throughout the writing of this book. We would also like to thank Dr. Alex Krizhevsky for gracefully giving us permission to use figures from the AlexNet paper. We would further like to acknowledge the developers of Theano and other Python libraries that are used by the yann toolbox and are used in the production of some of the figures in this book. In particular, we would like to thank Frédéric Bastien and Pascal Lamblin from the Theano users group and Montreal Institute of Machine Learning Algorithms of the Université de Montréal for the incredible customer support. We would also like to thank GitHub and Read the Docs for free online hosting of data, code, documentation, and tutorials.

Last, but foremost, we thank our friends and families for their unwavering support during this fun project and for their understanding and tolerance of many weekends and long nights spent on this book by the authors. We dedicate this book to them, with love.

**Ragav Venkatesan and Baoxin Li**

# Authors

**Ragav Venkatesan** is currently completing his PhD study in computer science in the School of Computing, Informatics and Decision Systems Engineering at Arizona State University (ASU), Tempe, Arizona. He has been a research associate with the Visual Representation and Processing Group at ASU and has worked as a teaching assistant for several graduate-level courses in machine learning, pattern recognition, video processing, and computer vision. Prior to this, he was a research assistant with the Image Processing and Applications Lab in the School of Electrical & Computer Engineering at ASU, where he obtained an MS degree in 2012. From 2013 to 2014, Venkatesan was with the Intel Corporation as a computer vision research intern working on technologies for autonomous vehicles. Venkatesan regularly serves as a reviewer for several peer-reviewed journals and conferences in machine learning and computer vision.

**Baoxin Li** received his PhD in electrical engineering from the University of Maryland, College Park, in 2000. He is currently a professor and chair of the Computer Science and Engineering program and a graduate faculty in the Electrical Engineering and Computer Engineering programs at Arizona State University, Tempe, Arizona. From 2000 to 2004, he was a senior researcher with SHARP Laboratories of America, Camas, Washington, where he was a technical lead in developing SHARP's trademarked HiMPACT Sports technologies. From 2003 to 2004, he was also an adjunct professor with Portland State University, Oregon. He holds 18 issued US patents and his current research interests include computer vision and pattern recognition, multimedia, social computing, machine learning, and assistive technologies. He won SHARP Laboratories' President's Award in 2001 and 2004. He also received the SHARP Laboratories' Inventor of the Year Award in 2002. He is a recipient of the National Science Foundation's CAREER Award.

# 1

# INTRODUCTION TO
# VISUAL COMPUTING

The goal of human scientific exploration is to advance human capabilities. We invented fire to cook food, therefore outgrowing our dependence on the basic food processing capability of our own stomach. This led to increased caloric consumption and perhaps sped up the growth of civilization—something that no other known species has accomplished. We invented the wheel and vehicles therefore our speed of travel does not have to be limited to the ambulatory speed of our legs. Indeed, we built airplanes, if for no other reason than to realize our dream of being able to take to the skies. The story of human invention and technological growth is a narrative of the human species endlessly outgrowing its own capabilities and therefore endlessly expanding its horizons and marching further into the future.

Much of these advances are credited to the wiring in the human brain. The human neural system and its capabilities are far-reaching and complicated. Humans enjoy a very intricate neural system capable of thought, emotion, reasoning, imagination, and philosophy. As scientists working on computer vision, perhaps we are a little tendentious when it comes to the significance of human vision, but for us, the most fascinating part of human capabilities, intelligence included, is the cognitive-visual system. Although human visual system and its associated cognitive decision-making processes are one of the fastest we know of, humans may not have the most powerful visual system among all the species, if, for example, acuity or night vision capabilities are concerned (Thorpe et al., 1996; Watamaniuk and Duchon, 1992). Also, humans peer through a very narrow range of the electromagnetic spectrum. There are many other species that have a wider visual sensory range than we do. Humans have also become prone to many corneal visual deficiencies such as near-sightedness. Given all this, it is only natural that we as humans want to work on improving our visual capabilities, like we did with other deficiencies in human capabilities.

We have been developing tools for many centuries trying to *see* further and beyond the eye that nature has bestowed upon us. Telescopes, binoculars, microscopes, and magnifiers were invented to see much farther and much smaller objects. Radio, infrared, and x-ray devices make us *see* in parts of the electromagnetic spectrum, beyond the visible band that we can naturally perceive. Recently, interferometers were perfected and built, extending human *vision* to include gravity waves, making way for yet another way to look at the world through gravitational astronomy. While all these devices extend the human visual capability, scholars and philosophers have long since realized that we do not see just with our eyes. Eyes are but mere imaging instruments; it is the brain that truly *sees*.

While many scholars from Plato, Aristotle, Charaka, and Euclid to Leonardo da Vinci studied *how* the eye sees the world, it was Hermann von Helmholtz in 1867 in his *Treatise on the Physiological Optics* who first postulated in scientific terms that the eye only captures images and it is the brain that truly *sees* and recognizes the objects in the image (Von Helmholtz, 1867). In his book, he presented novel theories on depth and color perception, motion perception, and also built upon da Vinci's earlier work. While it had been studied in some form or the other since ancient times in many civilizations, Helmholtz first described the idea of unconscious inference where he postulated that not all ideas, thoughts, and decisions that the brain makes are done so consciously. Helmholtz noted how susceptible humans are to optical illusions, famously quoting the misunderstanding of the sun revolving around the earth, while in reality it is the horizon that is moving, and that humans are drawn to emotions of a staged actor even though they are only staged. Using such analogies, Helmholtz proposed that the brain understands the images that the eye sees and it is the brain that makes inferences and understanding on what objects are being seen, without the person consciously noticing them. This was probably the first insight into neurological vision. Some early-modern scientists such as Campbell and Blakemore started arguing what is now an established fact: that there are neurons in the brain responsible for estimating object sizes and sensitivity to orientation (Blakemore and Campbell, 1969). Later studies during the same era discovered more complex intricacies of the human visual system and how we perceive and detect color, shapes, orientation, depth, and even objects (Field et al., 1993; McCollough, 1965; Campbell and Kulikowski, 1966; Burton, 1973).

The above brief historical accounts serve only to illustrate that the field of computer vision has its own place in the rich collection of stories of human technological development. This book focuses on a concise presentation of modern computer vision techniques, which might be stamped as *neural* computer vision since many of them stem from artificial neural networks. To ensure the book is self-contained, we start with a few foundational chapters that introduce a reader to the general field of visual computing by defining basic concepts, formulations, and methodologies, starting with a brief presentation of image representation in the subsequent section.

### Image Representation Basics

Any computer vision pipeline begins with an imaging system that captures light rays reflected from the scene and converts the optical light signals into an image in a format that a computer can read and process. During the early years of computational imaging, an image was obtained by digitizing a film or a printed picture; contemporarily, images are typically acquired directly by digital cameras that capture and store an image of a scene in terms of a set of ordered numbers called pixels. There are many textbooks covering image acquisition and a camera's inner workings (like its optics, mechanical controls and color filtering, etc.) (Jain, 1989; Gonzalez and Woods, 2002), and thus we will present only a brief account here. We use the simple illustration of Figure 1.1 to highlight the key process of sampling (i.e., discretization via the image grid) and quantization (i.e., representing each pixel's color values with only a finite set of integers) of the light ray coming from a scene into the camera to form an image of the world.

Practically any image can be viewed as a matrix (or three matrices if one prefers to explicitly consider the color planes separately) of quantized numbers of a certain bit length encoding the intensity and color information of the optical projection of a scene onto the imaging plane of the camera. Consider Figure 1.1. The picture shown was captured by a camera as follows: The camera has a sensor array that determines the

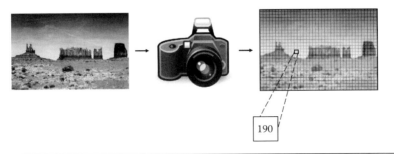

**Figure 1.1**    Image sampling and quantization.

size and resolution of the image. Let us suppose that the sensor array had $n \times m$ sensors, implying that the image it produced was $n \times m$ in its size. Each sensor grabbed a sample of light that was incident on that area of the sensor after it passed through a lens. The sensor assigned that sample a value between 0 and $(2^b - 1)$ for a $b$-bit image. Assuming that the image was 8 bit, the sample will be between 0 and 255, as shown in Figure 1.1. This process is called sampling and quantization, *sampling* because we only picked certain points in the continuous field of view and *quantization*, because we limited the values of light intensities within a finite number of choices. Sampling, quantization, and image formation in camera design and camera models are themselves a much broader topic and we recommend that interested readers follow up on the relevant literature for a deeper discussion (Gonzalez and Woods, 2002). Cameras for color images typically produce three images corresponding to the red (R), green (G), and blue (B) spectra, respectively. How these R, G, and B images are produced depends on the camera, although most consumer-grade cameras employ a color filter in front of a single sensor plane to capture a mosaicked image of all three color channels and then rely on a "de-mosaicking" process to create full-resolution, separate R, G, and B images.

With this apparatus, we are able to represent an image in the computer as stored digital data. This representation of the image is called the pixel representation of the image. Each image is a matrix or tensor of one (grayscale) or three (colored) or more (depth and other fields) channels. The ordering of the pixels is the same as that of the ordering of the samples that were collected, which is in turn the order of the sensor locations from which they were collected. The higher the value of the pixel, the greater the intensity of color present. This is the most explicit representation of an image that is possible. The larger the image, the more pixels we have. The closer the sensors are, the higher resolution the produced image will have when capturing details of a scene. If we consider two images of different sizes that sample the same area and field of view of the real world, the larger image has a higher *resolution* than the smaller one as the larger image can resolve more detail. For a grayscale image, we often use a two-dimensional discrete array $I(n_1, n_2)$ to represent the underlying matrix of pixel values, with $n_1$ and $n_2$ indexing the pixel at the $n_1^{th}$ row and the $n_1^{th}$ column of the matrix, and the value of $I(n_1, n_2)$ corresponding to the pixel's intensity, respectively.

While each pixel is sampled independently of the others, the pixel incenties are in general not independent of each other. This is because a typical scene does not change drastically everywhere and thus adjacent samples will in general be quite similar, except for pixels lying on the border between two visually different entities in the world. Therefore, *edges* in images that are defined by discontinuities (or large changes) in pixel values, are a good indicator of entities in the image. In general, images capturing a natural scene would be smooth (i.e., with no changes or only small changes) everywhere except for pixels corresponding to the edges.

The basic way of representing images as matrices of pixels as discussed above is often called spatial domain representation since the pixels are viewed as measurements, sampling the light intensities in the space or more precisely on the imaging plane. There are other ways of looking at or even acquiring the images using the so-called frequency-domain approaches, which decompose an image into its frequency components, much like a prism breaking down incident sunlight into different color bands. There are also approaches, like wavelet transform, that analyze/decompose an image using time–frequency transformations, where *time* actually refers to *space* in the case of images (Meyer, 1995). All of these may be called transform-domain representations for images. In general, a transform-domain representation of an image is invertible, meaning that it is possible to go back to the original image from its transform-domain representation. Practically, which representation to use is really an issue of convenience for a particular processing task. In addition to representations in the spatial and transform domains, many computer vision tasks actually first compute various types of *features* from an image (either the original image or some transform-domain representation), and then perform some analysis/inference tasks based on the computed features. In a sense, such computed features serve as a new representation of the underlying image, and hence we will call them feature representations. In the following section, we briefly introduce several commonly used transform-domain representations and feature representations for images.

*Transform-Domain Representations*

Perhaps the most-studied transform-domain representation for images (or in general for any sequential data) is through Fourier analysis (see Stein and Shakarchi, 2003). Fourier representations use linear combinations of sinusoids to represent signals. For a given image $I(n_1, n_2)$, we may *decompose* it using the following expression (which is the inverse Fourier transform):

$$I(n_1, n_2) = \frac{1}{nm} \sum_{u=0}^{n-1} \sum_{v=0}^{m-1} I_F(u, v) e^{j2\pi \left( \frac{un_1}{n} + \frac{vn_2}{m} \right)} \tag{1.1}$$

where, $I_F(u, v)$ are the Fourier coefficients and can be found by the following expression (which is the Fourier transform):

$$I_F(u, v) = \sum_{n_1=0}^{n-1} \sum_{n_2=0}^{m-1} I(n_1, n_2) e^{-j2\pi \left( \frac{un_1}{n} + \frac{vn_2}{m} \right)} \tag{1.2}$$

In this representation, the pixel representation of the image $I(n_1, n_2)$ is broken down into frequency components. Each frequency component has an associated coefficient that describes *how much* that frequency component is present. Each frequency component becomes the basis with which we may now represent the image. One popular use of this approach is the variant discrete cosine transform (DCT) for Joint Photographic Experts Group (JPEG) image compression. The JPEG codec uses only the cosine components of the sinusoid in Equation 1.2 and is therefore called the discrete cosine basis. The DCT basis functions are picturized in Figure 1.2.

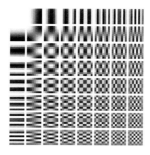

**Figure 1.2**   JPEG DCT basis functions.

Any kernel of a transform going from a pixel representation to a transform-domain representation and back can be written as $b(n_1,n_2,u,v)$ for going forward and $b'(n_1,n_2,u,v)$ for the inverse. For many transforms, often these bases are invertible under closed mathematical formulations to obtain one from the other. A projection or transformation from an image space to a basis space can be formulated as

$$I_T(u,v) = \sum_{n_1=0}^{n-1}\sum_{n_2=0}^{m-1} I(n_1,n_2)b(n_1,n_2,u,v) \tag{1.3}$$

and its inverse as,

$$I(n_1,n_2) = \sum_{u=0}^{n-1}\sum_{v=0}^{m-1} I_T(u,v)b'(n_1,n_2,u,v) \tag{1.4}$$

Equation 1.3 is a generalization of Equation 1.2. Many image representations can be modeled by this formalization, with Fourier transform being a special case.

*Image Histograms*

As the first example of feature representations for images, we discuss the histogram of an image, which is a global feature of the given image. We start the discussion by considering a much simpler feature defined by Equation (1.5):

$$I_m = \frac{1}{nm} \sum_{n_1=0}^{n-1} \sum_{n_2=0}^{m-1} I(n_1, n_2) \qquad (1.5)$$

This representation is simply the mean of all the pixel values in an image and is a scalar. We can have representations that are common for the whole image such as this. This is not very useful in practice for elaborate analysis tasks, since many images may have very similar (or even exactly the same) means or other image-level features. But a representation as simple as the mean of all pixels does provide some basic understanding of the image, such as whether the image is dark or bright, holistically speaking.

An 8-bit image has pixel values going from 0 to 255. By counting how many pixels are taking, respectively, one of the 256 values, we can obtain a distribution of pixel intensities defined on the interval [0, 255] (considering only integers), with the values of the function corresponding to the counts (or counts normalized by the total number of pixels in the image). Such a representation is called a histogram representation. For a color image, this can be easily extended to a three-dimensional histogram. Figure 1.3 shows an image and the corresponding histogram.

More formally, if we have a $b$-bit representation of the image, we have $2^b$ quantization levels for the pixel values; therefore the image is represented by values in the range 0 to $2^b - 1$. The (normalized) histogram representation of an image can now be represented by

$$I_b(i) = \frac{1}{nm} \sum_{n_1=0}^{n-1} \sum_{n_2=0}^{m-1} \mathbb{I}(I(n_1, n_2) = i), \, i \in [0, 1, \ldots, 2^b - 1] \qquad (1.6)$$

In Equation 1.6, $\mathbb{I}$ is an indicator function that takes the value 1 whenever its argument (viewed as a logic expression) is true, and 0 otherwise. The normalization of the equation with $\frac{1}{nm}$ is simply a way to help alleviate the dependence of this feature on the actual size of the image.

While histograms are certainly more informative than, for example, the mean representation defined earlier, they are still very

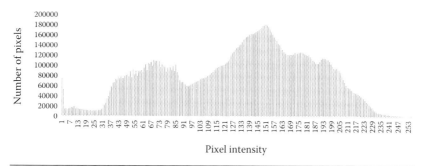

**Figure 1.3**   Image and its histogram.

coarse global features: Images with drastically different visual content might all lead to very similar histograms. For instance, many images of different patients whose eyes are retinal-scanned using a fundus camera can all have very similar histograms, although they can show drastically different pathologies (Chandakkar et al., 2013).

A histogram also loses all spatial information. For example, given one histogram indicating a lot of pixels in some shade of green, one cannot infer where those green pixels may occur in the image. But a trained person (or machine) might still be able to infer some global properties of the image. For instance, if we observe a lot of values on one particular shade of green, we might be able to infer that the image is outdoors and we expect to see a lot of grass in it. One might also go as far as to infer from the shade of green that the image might have been captured at a Scandinavian countryside.

Histogram representations are much more compact than the original images: For an 8-bit image, the histogram is effectively an array of 256 elements. Note that, one may further reduce the number of elements by using only, for example, 32 elements by further quantizing the grayscale levels (this trick is more often used in computing color histograms to make the number of distinctive color bins more manageable). This suggests that we can efficiently compare images in applications where the histogram can be a good feature. Normalization by the size of an image also makes the comparison between images of different sizes possible and meaningful.

*Image Gradients and Edges*

Many applications demand localized features since global feature representations like color histograms will not serve a useful purpose. For instance, if we are trying to program a computer to identify manufacturing defects in the microscopic image of a liquid-crystal display (LCD) (e.g., for automated quality control in LCD production), we might need to look for localized edge segments that deviate from the edge map of an image from a defect-free chip.

Detecting localized features like edges from an image is typically achieved by *spatial filtering* (or simply *filtering*) the image (Jain et al., 1995). If we are looking for a pattern, say a quick transition of pixels from a darker to a brighter region going horizontally left to right (a vertical rising edge), we may design a *filter* that, when applied to an image, will produce an image of the same size as the input but with a higher value for the pixels where this transition is present and a lower value for the transition being absent. The value of this response indicates the *strength* of the pattern that we are looking for. The implementation of the filtering process may be done by *convolution* of a template/mask (or simply a filter) with the underlying image $I(n_1, n_2)$. The process of convolution will be explained in detail later.

In the above example of looking for vertical edges, we may apply a simple one-dimensional mask of the form [−1, 0, 1] to all the rows of the image. This filter, when *convolved* with an image, would produce another image whose pixel values (considering only the magnitude, since the filtering results may contain negative numbers) indicate how dramatically the pixels around a given location *rise* or *fall* from left to right. It is evident that convolution using the above filter is equivalent to calculation of the differentiation of $I(n_1, n_2)$ along the horizontal direction. Hence, the filter is a gradient filter since its output is the horizontal gradient of the image $I(n_1, n_2)$. If one is interested in a binary edge map, where only the locations of strong transitions are kept, the above response image from the filter can be thresholded to produce that. Figure 1.4 demonstrates the use of the above simple filter by applying it on a quality control image from the LCD industry.

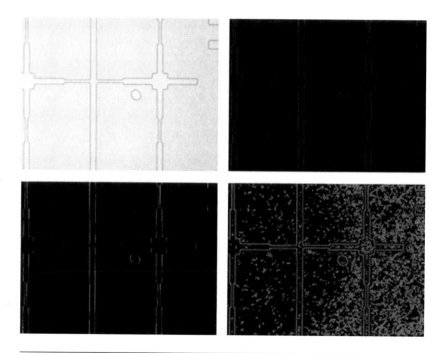

**Figure 1.4**    In the top row, the left is an image showing details of an LCD quality analysis photograph sample; on the right is the same image convolved with a [−1 0 1] filter. In the bottom row, the left image is the edge map threshold from the previous convolution and the last image is the output of a Canny edge detector, with spurious edges detected. Note how the one-dimensional edge detector does not detect horizontal edges but Canny detects those edges too.

**Table 1.1** Some Popular Edge Detection Filters

| OPERATOR | HORIZONTAL | VERTICAL | DIAGONAL | ANTIDIAGONAL |
|---|---|---|---|---|
| **Roberts** | $\begin{matrix}-1 & 0\\ 0 & 1\end{matrix}$ | $\begin{matrix}0 & -1\\ 1 & 0\end{matrix}$ | | |
| **Prewitt** | $\begin{matrix}-1 & -1 & -1\\ 0 & 0 & 0\\ 1 & 1 & 1\end{matrix}$ | $\begin{matrix}-1 & 0 & 1\\ -1 & 0 & 1\\ -1 & 0 & 1\end{matrix}$ | $\begin{matrix}0 & 1 & 1\\ -1 & 0 & 1\\ -1 & -1 & 0\end{matrix}$ | $\begin{matrix}-1 & -1 & 0\\ -1 & 0 & 1\\ 0 & 1 & 1\end{matrix}$ |
| **Sobel** | $\begin{matrix}-1 & -2 & -1\\ 0 & 0 & 0\\ 1 & 2 & 1\end{matrix}$ | $\begin{matrix}-1 & 0 & 1\\ -2 & 0 & 2\\ -1 & 0 & 1\end{matrix}$ | $\begin{matrix}0 & 1 & 2\\ -1 & 0 & 1\\ -2 & -1 & 0\end{matrix}$ | $\begin{matrix}-2 & -1 & 1\\ -1 & 0 & 1\\ 0 & 1 & 2\end{matrix}$ |

The above simple filter may be transposed to detect transitions in other directions. Further, typically two-dimensional filters are used for images, and the size of the filters may vary too.

Some of the earliest edge detection filters are summarized in Table 1.1. Although these rudimentary filters may produce image gradients that can be thresholded to form an edge map, one may get very noisy results from such a simplistic approach. For one thing, it is not easy to determine a good threshold. Also, the raw gradient values from a natural image may remain strong for a small neighborhood (as opposed to being large only along a thin line), and thus simple thresholding may lead to many thick edges (if they still give an impression of an edge at all). Therefore, some postprocessing steps are typically required to produce edge maps that are more localized (thin) and conform better to the perceived boundaries of the regions in the original image. Perhaps the best-known approach for a very good postprocessing job is the Canny edge detector (Canny, 1986), whose results are also included in Figure 1.4.

We now formalize a little bit the filtering or convolution process, which we have loosely described above. Consider a two-dimensional filter (or mask, or template, or kernel, all used interchangeably in this context), $F(l,w)$, defined on $[-a,a] \times [-b,b]$, we center the filter around the origin $0,0$ for convenience. Convolving the filter with an image $I(n_1,n_2)$ and producing an output image $Z(n_1,n_2)$ is represented in the following:

$$Z(n_1,n_2) = \sum_{l=-a}^{a}\sum_{w=-b}^{b} F(l,w)I(n_1+l,n_2+w)\forall n_1,n_2 \qquad (1.7)$$

Note that the above convolution is supposed to produce an output of the same size as the input. Some care needs to be taken when dealing with the pixels on the boundaries of the image so that the convolution is well supported. Typical ways of doing this include padding as many additional rows/columns as necessary along the borders before doing the convolution to maintain the size of the image. This type of a convolution is called a 'same' convolution. If padding were avoided, we get a "valid" convolution whose size is smaller than the original image by $(a-1) \times (b-1)$ pixels. The two-dimensional filter $F$ itself can be considered a small image of size $(2a+1) \times (2b+1)$. Although square filters are most common, we intentionally use different sizes ($a$ and $b$) for the horizontal and vertical dimensions to indicate the possibility of allowing nonsquare filters.

The gradient filters discussed thus far are based on first-order differentiations of the image $I(n_1, n_2)$. Laplacians are second-order filters, which will detect a ramp-like edge as a pair of positive and negative peaks in the second-order derivatives and hence help localization of the edge as the *zero-crossing* between the two peaks. Mathematically, a Laplacian of a two-dimensional function $f(x, y)$ is defined by

$$f_L'' = \frac{\partial^2 f}{\partial x^2} + \frac{\partial^2 f}{\partial y^2} \tag{1.8}$$

Masks implementing approximations of the Laplacians are shown in Table 1.2. The Laplacian is isotropic. Also, it is susceptible to noise due to its second-order nature. Hence, it is often applied on an image that has been smoothed by some filtering process for noise reduction.

Smoothing an image may be achieved by averaging pixels in a small neighborhood. A simple averaging filter is more commonly referred to as a box filter, which is of the form:

$$B(u, v) = \frac{1}{U \times V} 1 \ \forall u \in [1, 2, \ldots U], v \in [1, 2, \ldots, V] \tag{1.9}$$

**Table 1.2**  Laplacian Operators

| 0  | −1 | 0  | −1 | −1 | −1 |
|----|----|----|----|----|----|
| −1 | 4  | −1 | −1 | 8  | −1 |
| 0  | −1 | 0  | −1 | −1 | −1 |

where $U \times V$ is the size of the filter. For instance, a $3 \times 3$ averaging box filter is of the form $\frac{1}{9} \begin{bmatrix} 1 & 1 & 1 \\ 1 & 1 & 1 \\ 1 & 1 & 1 \end{bmatrix}$. This filter would average pixels from a neighborhood of nine pixels. A Gaussian smoothing filter is a weighted average that weighs the pixels proportionally using a Gaussian function: The farther the point is from the center, the lower it is weighted. Gaussian filters are of the form:

$$G(u,v) = \frac{1}{2\pi\sigma^2} e^{\left(-\frac{u^2+v^2}{\sigma^2}\right)} \tag{1.10}$$

Apart from giving us a weighted average useful in noise reduction, Gaussian smoothing filters are in general preferred over box filters since a Gaussian-shaped kernel is often used to model defocus and the point spread function of a sensor unit on the imaging plane. Gaussians can also be approximated into 2D kernels to be used as convolutional masks. A typical approximation of a $3 \times 3$ Gaussian is $\frac{1}{16} \begin{bmatrix} 1 & 2 & 1 \\ 2 & 4 & 2 \\ 1 & 2 & 1 \end{bmatrix}$. The advantage with Gaussians is that the Fourier representation of a Gaussian is another Gaussian. With the use of the convolution theorem to great effect, we can create faster and more efficient convolutions if we use Gaussians.

The Laplacian of the Gaussian (LoG) is the Laplacian applied to a Gaussian-filtered image. The Gaussian filter smooths the image and the Laplacian detects the zero-crossings on a smoothed image. The points that it detects are usually good key points around which there might lurk good attributes. Since the convolution operation is associative, we can create a LoG filter independent of the image. The LoG filter takes the form:

$$\text{LoG}(u,v) = -\frac{1}{\pi\sigma^4}\left[1 - \frac{u^2 + v^2}{2\sigma^2}\right] e^{-\frac{u^2+v^2}{2\sigma^2}}, \tag{1.11}$$

which is essentially a second-order differential of the Gaussian function itself. The LoG can also be approximated by a convolutional

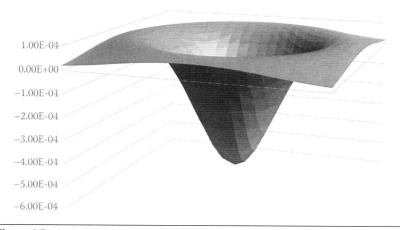

**Figure 1.5** Laplacian of Gaussian.

mask. This approximation can then be used as a mask to convolve with an image just like a normal convolution. Figure 1.5 is an example of a LoG mask.

### Going beyond Image Gradients

Image gradients and edges are very rudimentary image feature representations. Such features may already support some inference tasks. For example, the edge map of a template image of a certain object may be compared with the edge map of a given image to detect and locate the occurrence of that object in the given image, as done in Li et al. (2001b). However, these rudimentary features find more use in the construction of more sophisticated features, such as lines from the Hough transform (Hough, 1962), Harris corners (Harris and Stephens, 1988), the scale-invariant feature transform (SIFT) features (Lowe, 1999), and the histogram of oriented gradient (HOG) feature (Dalal and Triggs, 2005), which we discuss below only briefly since there are abundant references on them in the literature and omitting the details of their inner workings will not hinder our subsequent discussions on neural computer vision.

### Line Detection Using the Hough Transform

Edges detected from a natural image are often fragmented since most approaches to edge detection process the pixels by considering only a local neighborhood. If, for example, there is a building with straight lines as its contour in the image, an edge detector may not be able to produce a nicely connected contour, but rather some fragmented edge segments on the contour. Further, the edge map is nonparametric in the sense that one does not obtain any parametric model for the contour in the above example. The basic Hough transform for line detection is essentially a voting procedure in the two-dimensional parameter space for lines: any point in this space defines an equation for a line in the image domain, and thus any edge pixel casts a vote for any lines passing through that edge pixel. After all the votes are counted, local maxima in the parameter space define equations for lines that cover many edge pixels. While the basic version of the Hough transform is only for line detection, extensions have been made for detecting other shapes (Ballard, 1981).

### Harris Corners

Sometimes point features are desired. For example, in estimating global transformation between two images, one might need to find a small set of matched points in the given two images. A good feature point should have some uniqueness if it is to be useful. Local features like pixels with large spatial gradients may be found by gradient filters as discussed above, but in practice, many such features may be too similar to each other to be useful for a matching task. For example, pixels along the same edge segment might all have similar gradients and thus cannot be easily distinguished from each other when considered in isolation. Looking at the problem from another angle, if we examine a straight line/edge through only a small window and if the line/edge is moving along its axis, we may not even be able to notice the movement (until perhaps when the end of the line moves into the visible window). This is the so-called *aperture problem* in motion estimation in computer vision. The Harris corners are defined based on a principle derived from the above observation: a good point feature should be one whose motion, whichever direction it is, would be evident even if viewed only through a small window. Visually, such good point features typically correspond to some sort of "corners" in the image.

In implementation, corners are detected through Eigen analysis of the structure tensor, a matrix derived from local image gradients.

### Scale-Invariant Feature Transform

For a feature point to be practically useful, it should have some level of *invariance* to typical imaging or other conditions that impact the quality or appearance of the acquired images. For example, if we rely on some point features to match or track the same object across different images captured by different cameras from different viewpoints, we require that such features be invariant to changes in scale, rotation, and even appearance.

SIFT is an algorithm for detecting feature points (customarily called SIFT features) that are supposedly invariant to changes in scale and rotation, or have a slight appearance change due to varying illumination or local geometric distortion. The algorithm first creates a multiscale representation of the original image. Then it finds the extrema in the difference of Gaussians in that representation and uses them as potential feature points. Some criteria are used to discard any potential feature point that is deemed a poor candidate (e.g., low-contrast points). Finally, a descriptor will be computed for any of the remaining points and its neighboring region. This descriptor (and its location in an image) is basically what we call a SIFT feature. The full algorithm as described in Lowe (1999) also covers other implementation issues related to fast matching of objects based on SIFT features.

### Histogram of Oriented Gradients

So far we have seen both local and global feature representations of an image. Local histograms could represent the histogram of a patch of $k \times k, k < n, m$. This would describe the color or pixel distribution or edge distribution over a small region (Park et al., 2000). One popular locally global feature space is the HOGs first proposed by Dalal and Triggs (2005). Given an image we can easily obtain gradient magnitudes for all pixels in the image. Consider also quantizing the gradient directions into only $j$ possibilities/bins. Using a nonoverlapping sliding window of $k \times k$, we can run through the entire image and

 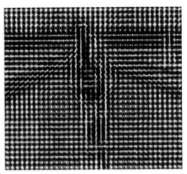

**Figure 1.6**   Visualization of HOG features. Figure generated using the Hoggles tool. (From Vondrick, Carl et al., Hoggles: Visualizing object detection features. *Proceedings of the IEEE International Conference on Computer Vision*, pp. 1–8, 2013).

create patchwise histograms, with the value of the histogram being the magnitude of the gradient in the corresponding direction. We will have a $j$-bin histogram for each patch, with each bin (representing some range of angles) holding the value of the magnitude of that direction's gradient in that block. The HOG feature is a shape representation of the image as some structural information is retained in the feature. Figure 1.6 shows a visualization of this representation. In this representation for each block, the length of the arrow describes the magnitude of the gradient bin in the direction that arrow is pointing to. Each block has an 8-bin histogram of gradients. As can be noticed, this representation shows the direction of edges in a manner that is better than trivial edge features. These show the orientation of gradients and also the strength of those gradients.

Thus far, we have introduced various types of filters that are designed to extract different features. These examples illustrate that filters can be designed to be anything that is feasible and that can effectively visualize an image in terms of some representation. A feature is after all a particular attribute with which we want to measure the data. All we need to do is to design good filters for computing good feature representations. Indeed, until convolutional neural networks (CNNs) became popular, a major task in computer vision was the designing of efficient and descriptive task-specific feature representations. Even today, good features are the main reason for the incredible performance of computer vision. Modern neural computer vision relies on neural networks in designing problem-specific features. Neural networks that design their own image features are only recently popular and it is helpful to study and understand how feature representations and machine learning worked prior to neural feature representations, before we proceed to machines that learn their own features.

*Decision-Making in a Hand-Crafted Feature Space*

We have reviewed some of the most common methods for representing images in spatial, transform, and feature domains. All these representations have found use in various computer vision applications. Even the basic matrix representation in the pixel domain can be used in simple tasks like template matching, change detection in video, region segmentations, etc. However, more complex tasks often demand elaborate processing steps that typically operate in some transform domain or some feature space. Consider, for example, object recognition in images. In general, we need some feature representation for the images to be processed (including feature-based models for the objects to be recognized), and then some decision-making procedures are employed to determine whether a new image should be classified as one of the given objects (or declared as not belonging to the given set of objects). The decision-making process is often based on a learning approach, where the parameters that control decision-making are figured out by using a *training set* of data samples (typically with known labeling information). Accordingly, machine learning approaches have been among the most used techniques in many computer vision tasks. This section will present basic approaches to decision-making in general computer vision problems where statistical learning approaches are employed to infer information from data. We will start with a trivial, but instructive, example of determining whether an image was captured during the day or night, using only the image (i.e., not relying on metadata that might have stored the time of day).

Suppose we have a dataset of both day and nighttime images and that we already know which images were captured at night or during the day (i.e., the images have been *labeled*). We further assume that these images were captured outdoors (and hence it is not the case where artificial lights control the image brightness).

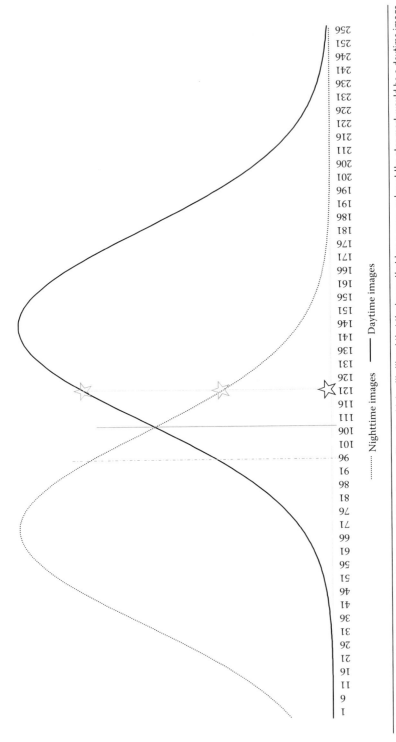

**Figure 1.7** Deciding between a daytime or a nighttime image. There is a higher likelihood that the image that has a mean value at the star mark would be a daytime image rather than a nighttime image. The broken line is the decision boundary with prior and the solid line is the decision boundary without the prior.

Let us extract a very simple feature representation for each image: The mean of all the pixel values in the images, such as the one described in Equation 1.5. Once all the means from the images are extracted, we can make a histogram plot of these mean values. Let us also make another assumption that the histogram plots can be modeled using Gaussian probability distributions, one for day and one for nighttime images. Figure 1.7 illustrates such distributions. The horizontal axis is all possible mean values from 0 to 255 and the vertical axis shows the relative counts of the number of images having a particular mean value. The dotted line is the histogram of the images that we know are nighttime images and the solid line is the histogram of the images that we know are daytime images. From this histogram, we see that nighttime images are generally darker than the daytime images, which should be intuitive. Furthermore, the histograms effectively give us an estimation of the probability of the mean value of the image taking some particular value. For example, we may say that the nighttime images are very likely to have an image mean around 64 (i.e., around the peak of the dotted line distribution), while the daytime images are more likely to have a mean around 145.

Now suppose that we are given a new image, whose capture time is unknown to us, and we want to estimate that. We compute the mean of this image and found it is 121 (shown by the *star* mark in Figure 1.7). Using the knowledge of the distributions that we have established from the training set, we can conclude that the new image is more likely a bright image taken during daytime than otherwise. The reason why we settled on this decision is because, at that mean value of 121, we have in our training dataset more than twice as many daytime images as nighttime images.

*Bayesian Decision-Making*

We now expand the above intuition to formalize a decision-making procedure: Bayesian classification. Probabilistically, Figure 1.7 is a tale of two Gaussian distributions. Let us index the images (represented by their means) by variable $i$ and thus we can use "image $i$" or simple $i$ to stand for the $i$th image in our dataset. Let us have a variable $\omega$ representing $i$ being a daytime or nighttime image. If

$\omega = 0$, the image is a nighttime image and thus $\omega = 1$ means that the image was shot during the day. Now the histograms in Figure 1.7 can be interpreted as the class-conditional probability density functions (PDFs): the dotted line represents the PDF of the image mean for nighttime images, that is, $p(x \mid \omega = 0)$, and similarly the solid line is for $p(x \mid \omega = 1)$. We call $p(x \mid \omega = 0)$ and $p(x \mid \omega = 1)$ class-conditional PDFs (or simply class-conditionals in short). We will further assume that each image is equally likely to be from daytime or nighttime (also called as the uninformative prior assumption). Or, from the given training set, if we see an equal number of images from either group, we will draw this conclusion. Formally, this is to say the *prior* probabilities of the two classes are equal: $P(\omega = 1) = P(\omega = 0) = 0.5$. Note that, in general, these prior probabilities may not be equal.

Given an image $j$, its mean X can be computed. The optimal (minimum average error) decision is to classify the image (as a daytime or nighttime image) by comparing the *a posteriori* probabilities: $P(\omega = 0 \mid X = x)$ and $P(\omega = 1 \mid X = x)$. Image $j$ will be given the class label (0 or 1) that gives rise to a larger *a posteriori* probability. Since the *a posteriori* probabilities can be related to the class-conditional PDFs $p(x \mid \omega)$ and the priors, the above decision rule can also be stated in terms of these quantities as

$$\omega = 0 \text{ if } P(\omega = 0)\, p(x \mid \omega = 0) > P(\omega = 1)\, p(x \mid \omega = 1) \text{ and}$$

$$\omega = 1 \text{ otherwise} \tag{1.12}$$

The incorporation of *priors* into this decision is central to Bayesian approaches. In the above example, if, say, we know from our dataset that, irrespective of the means, there are more images that are naturally taken during daytime than those that are taken during night, potentially because more people are awake and clicking during daytime and therefore are more likely to take images during daytime, that information is incorporated in the decision rule of Equation 1.12. Note that, when the feature value $x$ is given, the class-conditionals $p(x \mid \omega)$ become a measure of how likely that particular value $x$ comes from the underlying classes, and thus the class-conditionals are now viewed as the *likelihood functions* (of the classes, assuming a given $x$). In this view, a prior is the bias (based on prior knowledge we had) that we use to weigh the likelihood (based on what we can learn from the current data) in order to strike a balance between prior information and the current information.

In the above example, the decision rule will eventually boil down to finding some $x_d$, which forms a *decision boundary*, or simply a single point in this one-dimensional feature space. Any image with mean $X < x_d$ will be classified as a nighttime image and classified as a daytime image otherwise. In Figure 1.7, the decision boundary around $x_d = 109$ would be the optimal solution when the priors are equal, whereas $x_d = 95$ illustrates the optimal decision for a case where $P(\omega = 1) > 0.5$.

The Bayesian decision process is applicable whenever we are able to model the data in a feature space and the distributions (the class-conditionals) of the classes and the priors can somehow be obtained. In that case, optimal decision boundaries can be derived as above. In practice, both the priors and the class-conditionals need to be estimated from some training data. The priors are scalars and thus may be easily estimated by relative frequencies of the samples from each class. There are two general types of density estimation techniques: parametric and nonparametric. In the earlier example, we essentially assumed the PDFs of the image means were Gaussian (i.e., a parametric approach). A reader interested in density estimation may refer to standard textbooks like Duda et al. (2012).

*Decision-Making with Linear Decision Boundaries*

While Bayesian decision-making is elegant and optimal (in the sense of providing minimum average misclassification error), as long as all the necessary probabilities and densities are available, for many real applications, we cannot reliably estimate those required quantities. We may then set out to find some simple linear hyperplanes that directly separate the classes in the feature space. The linearity is introduced really only for simplifying the problem and the ideal/optimal class boundaries are of course, in general, not linear. In the earlier example, the problem becomes one of finding a single point $x_d$ (without relying on the densities). If we had a way of finding out that point, we could start making decisions. In a linear approach, this point is computed as a linear combination of the features. For example, one might use the following linear approach to find a possible $x_d$:

$$x_d = \frac{1}{2}\left( \frac{1}{N_1}\sum_{i=1}^{N_1} x_i + \frac{1}{N_2}\sum_{i=1}^{N_2} x_i \right) \qquad (1.13)$$

where $N_1$ and $N_2$ are the number of samples from class 1 and class 2, respectively. Intuitively, this simple linear estimate for the decision threshold $x_d$ is just the middle point between the two means computed respectively for each class of images. While there is no reason to believe this solution would be optimal (as opposed to the guaranteed optimality of the Bayesian decision approach), we enjoy the simplicity arising from the assumption of linearity: We effectively only did some simple additions of the samples (along with three divisions) and we obtained the solution without worrying about anything related to the underlying densities or probabilities. In practice, since we cannot guarantee the above solution is optimal, we may define a more general linear form of the solution as

$$x_d = \sum_{i=1}^{N_1+N_2} w_i x_i \qquad (1.14)$$

and then try to figure out the weights $w_i$, so as to obtain a better solution. It is obvious that the earlier solution of Equation 1.13 is a special case of the solution in Equation 1.14.

The above basic idea can be extended to higher dimensional feature space: For two-class classification, we will have a line as the decision boundary in a two-dimensional space, a plane in a three-dimensional space, and in general a hyperplane in higher dimensional spaces. The task is now to estimate the parameters for the hyperplane. There are different techniques to do this, and most of them rely on some optimization procedure: An objective function is first defined (which is a function of the hyperplane parameters), and then the optimal parameters are sought to minimize (or maximize, depending on the formulation) the objective function. There are both iterative approaches (e.g., those employing gradient descent) and one-step approaches (e.g., pseudoinverse used with minimum mean-squared error as the objective function).

Among all the linear approaches, perhaps the most well known is the support vector machine (SVM), where the so-called *margin*, the distance between the decision hyperplane and the closest samples, is maximized (Cortes and Vapnik, 1995; Chapelle et al., 1999). Nonlinear extensions to the basic linear approaches exist. The idea is to allow the original feature space to be (nonlinearly) mapped to a new space and then apply the linear approach in the new space, hence effectively achieving nonlinear decision boundaries in the original space. A kernel SVM attains this goal in a smart way by using the *kernel trick* and hence explicit feature mapping is not necessary (Bishop, 2001).

## A Case Study with Deformable Part Models

The various representations and the decision-making techniques we have discussed so far seem to have equipped us with necessary gears for attacking any computer vision problem. Unfortunately, real applications may be much more complicated than the illustrative examples we have seen so far. Consider, for example, the task of recognizing nonrigid objects. Even those advanced features we introduced earlier may fall short of providing adequate description of the objects under varying structures and poses, let alone supporting the deployment of a decision-making algorithm like SVM. We now use this task of recognizing nonrigid objects, such as humans in an image, as a case study for illustrating how to develop additional approaches, building on top of the basic representations/tools we have introduced thus far.

Consider HOG features as previously discussed. If we want to detect pedestrians on roads, for example, we can create a dataset of images that have pedestrians with bounding boxes on them, and images that either have pedestrians where the bounding boxes are not on them or have non-human objects boxed. This accounts for our dataset from which we want to learn what a pedestrian on the road is and what is not. We then extract HOG features for all the bounding boxes such as the one in Figure 1.6. We can use these features to train a classifier, using one of the statistical learning techniques described earlier or maybe even a simple template matching scheme.

In doing so, we are expecting the system to learn to detect objects that are of a certain structure in certain poses and these poses only. While this may work in some cases such as upright humans, clearly there are cases where this will be insufficient. Deformable part models (DPMs) create models for objects with the assumption that models are made of subparts. These were proposed several times in the past with modifications and revisions under the name *pictorial structures* (Fischler and Eschlager, 1973; Felzenszwalb and Huttenlocher, 2000). The subparts are detected as objects as previously and the main object is built using the subparts. While we only know from our dataset a bounding box for where the object is in the given image, we do not know what their parts are. These parts and their relative locations are *learnt* as part of the algorithm itself.

DPMs work using two components: the parts and the springs. Parts are HOG templates for each of the parts themselves, and springs are a geometric prior or a cost of having a part at a certain location relative to the other parts. Part models create a matching score for the part themselves and the spring ensures that the object as a whole can be represented in a meaningful fashion using the parts. This aligns in a sense with the so-called Gestalt principles that deal with reasoning about the whole from the parts (e.g., see Tuck, 2010). This enables us to model a lot of varied poses and perspectives.

There have been several gradual updates to DPMs. The most recent incarnation, which uses only modeling and not grammar, was one

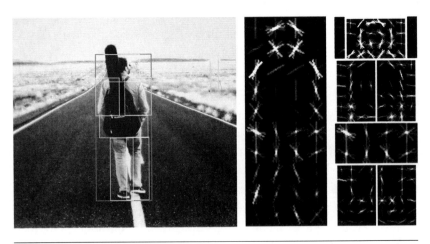

**Figure 1.8**   Deformable part models. On the left, a person is detected using the root filter and the part filter, in the middle is a person's root filter, and on the right are the part filters. This picture was produced using the DPM VOC release 1 code. (From Felzenszwalb, Pedro F. et al., Cascade object detection with deformable part models. *Computer Vision and Pattern Recognition (CVPR), 2010 IEEE Conference on.* IEEE. pp. 2241–2248, 2010.)

of the popular computer vision systems until deep CNNs took over (Felzenszwalb et al., 2010). DPM creates a score for each object and its location. The score is a combination of the root score, which is exactly the same as the Dalal and Triggs method, and the scores of all part matches and the scores of all the springs (Dalal and Triggs, 2005). It can be described using Equation (1.15):

$$s(p_1, p_2, \ldots, p_n) = \sum_{i=1}^{n} m_i(p_i) - \sum_{(i,j) \in E} d_{ij}(p_i, p_j) \qquad (1.15)$$

The first term is the matching score for the parts and the second score is the sum of pairwise spring scores for all parts relative to each other. If this score is maximized over all combinations of parts, $p_1, p_2, \ldots, p_n$, we get a hit in matching. Some implementations of this algorithm use dynamic programming to search through all of these parts and have been released for public use. Figure 1.8 illustrates an example using one such publicly available package. Modeling objects like this is more reliable than using only root models and thus has led to higher performance in many state-of-the-art object detection benchmark datasets.

**Migration toward Neural Computer Vision**

Thus far, we have studied some basics of what form the fundamental building blocks of traditional computer vision. As we see from the preceding examples, a traditional computer vision pipeline involves defining some feature representation, be it local or global or some combined representation of local and global properties, and a pattern detector/classifier, be it Bayesian or SVMs or perhaps model-based pattern detectors such as deformable part models. Such a system is illustrated in Figure 1.9.

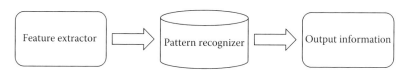

**Figure 1.9**   Traditional computer vision system.

Obviously, in such a processing pipeline, the innovation opportunities for performance improvement lie in finding a better feature representation, or developing a better classification scheme, or often both since a new feature representation often demands some new algorithms for classification. It is often challenging to rely on any explicit set of rules for designing perfect features and/or classifiers for a given problem, but on the other hand, some sort of training data may be available, and many computer vision researchers have been employing machine learning approaches to address the need for new methods. For example, in Nagesh and Li (2009), a new representation for face recognition was proposed, leveraging the compressive sensing theory (Candes et al., 2006) in compactly representing a set of images from the same subject. In Kulkarni and Li (2011), image features are defined as affine sparse codes that are learned from a large collection of image patches, and such features are then used for image categorization. An example of joint feature and classier learning is Zhang and Li (2010), where a dictionary for sparse coding (i.e., a feature representation) as well as a classifier using the sparse representation of face images under the dictionary is jointly learned from some labeled face images. It is worth noting that many manifold learning techniques seek more efficient representations of the data in some subspaces where features for classification may be done better, and thus they directly contribute to the feature extraction stage of the above processing pipeline (Belkin and Niyogi, 2001; Tenenbaum et al., 2000; Roweis and Saul, 2000).

While progress has been made, a pipeline such as that in Figure 1.9 is not without issues. One prominent difficulty is that the explicit design of a feature extractor is, in most cases, application and data dependent and thus a designer would often need to handcraft new features for a new application or even a new type of data for the same existing application. Also, handcrafting features is, by itself, a task requiring a lot of creativity, domain knowledge and experience from trial and error, and thus it is unrealistic to expect anyone to just come up with something close to an optimal solution (if one exists) (Venkatesan, 2012).

Recent years have witnessed the resurfacing of a class of approaches employing CNNs for addressing the above challenge, many under the name of *deep learning*. Artificial deep neural networks refer to a class of learning approaches where the weights for an input–output mapping are structured according to a network (LeCun, 2015). The key task of learning the weights is in fact quite similar to what we introduced earlier for learning the weights of linear classifiers, although neural networks typically learn hierarchical and nonlinear mappings. We have intentionally delayed the discussion of this class of learning approaches since CNNs and their variants are a focus of this book, and we will elaborate further in subsequent chapters.

Earlier in this chapter we saw examples of a kernel/mask being used to convolve with an image for gradient computation. CNNs have classification layers and convolutional layers, and the latter employs convolutional kernels whose weights are learned from the data, thus effectively achieving feature learning and classifier learning within the same framework. Using a CNN of many layers was already found long ago to be able to deliver the best performance for a challenging automatic target recognition task (Li et al., 1998, 2001a), and using learned kernels as general features was also attempted before (Li and Li, 1999). But nowadays the availability of massive datasets as well as a phenomenal increase of computing power (compared with 15–20 years back) have enabled the training of much deeper networks for much more challenging problems. It appears that neural computer vision has now arrived in an era in favor of a data-driven approach to learning not only classifier but also feature representations, and indeed some CNN models have been used as off-the-shelf feature detectors (Krizhevsky et al., 2012; Simonyan and Zisserman, 2014; Szegedy et al., 2015; Soekhoe et al., 2016; LeCun, 2015). This concise book will walk the reader through an exploration of modern neural computer vision approaches employing deep networks. We will start by presenting the necessary primaries in next couple of chapters.

## Summary

The history of humanity is measured by the constant increase of human capabilities by the development of technology. In this quest, computer vision has been at the forefront to help us *see* better in all contexts. While handcrafted features have been popular, recent advances and the popularity of deep neural network-based learning have shown much promise and potential. The book will discuss the new class of approaches built with CNNs that has greatly contributed to many recent breakthroughs in computer vision.

# References

Ballard, Dana H. 1981. Generalizing the Hough transform to detect arbitrary shapes. *Pattern Recognition* (Elsevier) 13: 111–122.

Belkin, Mikhail and Niyogi, Partha. 2001. Laplacian eigenmaps and spectral techniques for embedding and clustering. *NIPS Proceedings of the 14th International Conference on Neural Information Processing Systems*, 585–591, Vancouver, British Columbia, Canada.

Bishop, Christopher M. 2001. *Bishop Pattern Recognition and Machine Learning*. New York, NY: Springer.

Blakemore, Colin and Campbell, Frederick W. 1969. On the existence of neurones in the human visual system selectively sensitive to the orientation and size of retinal images. *Journal of Physiology* 203: 237.

Burton, Greg J. 1973. Evidence for non-linear response processes in the human visual system from measurements on the thresholds of spatial beat frequencies. *Vision Research* (Elsevier) 13: 1211–1225.

Campbell, Frederick W and Kulikowski, Janus J. 1966. Orientational selectivity of the human visual system. *Journal of Physiology* 187: 437.

Candes, Emmanuel J, Romberg, Justin K, and Tao, Terence. 2006. *Stable Signal Recovery from Incomplete and Inaccurate Measurements*. Hoboken, NJ: Wiley Online Library. 1207–1223.

Canny, John. 1986. A computational approach to edge detection. *Transactions on Pattern Analysis and Machine Intelligence* (IEEE) 8: 679–698.

Chandakkar, Parag S, Venkatesan, Ragav, and Li, Baoxin. 2013. Retrieving clinically relevant diabetic retinopathy images using a multi-class multiple-instance framework. *SPIE Medical Imaging. International Society for Optics and Photonics.* 86700Q, Orlando, Florida.

Chapelle, Olivier, Haffner, Patrick, and Vapnik, Vladimir N. 1999. Support vector machines for histogram-based image classification. *IEEE Transactions on Neural Networks* (IEEE) 10: 1055–1064.

Cortes, Corinna and Vapnik, Vladimir. 1995. Support-vector networks. *Machine Learning* (Springer) 20: 273–297.

Dalal, Navneet and Triggs, Bill. 2005. Histograms of oriented gradients for human detection. *IEEE Computer Society Conference on Computer Vision and Pattern Recognition* (CVPR'05). IEEE, pp. 886–893, San Diego, California.

Duda, Richard O, Hart, Peter E, and Stork, David G. 2012. *Pattern Classification*. Hoboken, NJ: John Wiley & Sons.

Felzenszwalb, Pedro F, Girshick, Ross B, and McAllester, David. 2010. Cascade object detection with deformable part models. *2010 IEEE Conference on Computer Vision and Pattern Recognition (CVPR).* IEEE, pp. 2241–2248, San Francisco, California.

Felzenszwalb, Pedro F and Huttenlocher, Daniel P. 2000. Efficient matching of pictorial structures. *IEEE Conference on Computer Vision and Pattern Recognition, 2000. Proceedings.* IEEE, pp. 66–73, Hilton Head, South Carolina.

Field, David J, Hayes, Anthony, and Hess, Robert F. 1993. Contour integration by the human visual system: Evidence for a local "association field". *Vision Research* (Elsevier) 33: 173–193.

Fischler, Martina A and Eschlager, Roberta A. 1973. The representation and matching of pictorial patterns. *IEEE Transactions on Computers* (IEEE) 22: 67–92.

Gonzalez, Rafael C and Woods, Richard E. 2002. *Digital Image Processing.* Upper Saddle River, NJ: Prentice Hall.

Harris, Chris and Stephens, Mike. 1988. A combined corner and edge detector. *Alvey Vision Conference* (Citeseer) 15: 50.

Hough, Paul VC. 1962. Method and means for recognizing complex patterns. US Patent 3,069,654.

Huo, Yuan-Kai, Wei, Gen, Zhang, Yu-Dong et al. 2010. An adaptive threshold for the Canny Operator of edge detection. *International Conference on Image Analysis and Signal Processing.* IEEE, pp. 371–374, Dallas, Texas.

Jain, Anil K. 1989. *Fundamentals of Digital Image Processing.* Upper Saddle River, NJ: Prentice-Hall.

Jain, Ramesh, Kasturi, Rangachar, and Schunck, Brian G. 1995. *Machine Vision* (Vol. 5). New York, NY: McGraw-Hill.

Krizhevsky, Alex, Sutskever, Ilya, and Hinton, Geoffrey E. 2012. ImageNet classification with deep convolutional neural networks. *Advances in Neural Information Processing Systems.* Harrahs and Harveys, Lake Tahoe, NIPS, pp. 1097–1105.

Kulkarni, Naveen and Li, Baoxin. 2011. Discriminative affine sparse codes for image classification. *IEEE Conference on Computer Vision and Pattern Recognition (CVPR), 2011.* IEEE, 1609–1616, Colorado Springs, Colorado..

LeCun, Yann. 2015. It's learning all the way down. *IEEE International Conference on Computer Vision.* Santiago.

Li, Baoxin, Chellappa, Rama, Zheng, Qinfen et al. 2001a. Empirical evaluation of FLIR-ATR algorithms–A comparative study. *Computer Vision and Image Understanding* (Elsevier) 84: 5–24.

Li, Baoxin, Chellappa, Rama, Zheng, Qinfen et al. 2001b. Model-based temporal object verification using video. *IEEE Transactions on Image Processing* (IEEE) 10: 897–908.

Li, Bao-Qing and Li, Baoxin. 1999. Building pattern classifiers using convolutional neural networks. *IEEE International Joint Conference on Neural Networks* 5: 3081–3085.

Li, Baoxin, Zheng, Qinfen, Der, Sandor Z et al. 1998. Experimental evaluation of neural, statistical, and model-based approaches to FLIR ATR. *International Society for Optics and Photonics Aerospace/Defense Sensing and Controls.* 388–397.

Lowe, David G. 1999. Object recognition from local scale-invariant features. *ICCV '99 Proceedings of the International Conference on Computer Vision.* IEEE, pp. 1150–1157, Fort Collins, Colorado.

McCollough, Celeste. 1965. Color adaptation of edge-detectors in the human visual system. *Science* (American Association for the Advancement of Science) 149: 1115–1116.

Meyer, Yves. 1995. *Wavelets and Operators.* Cambridge: Cambridge University Press.

Nagesh, Pradeep and Li, Baoxin. 2009. A compressive sensing approach for expression-invariant face recognition. *IEEE Conference on Computer Vision and Pattern Recognition, 2009. CVPR 2009.* IEEE, pp. 1518–1525, Miami, Florida.

Park, Dong Kwon, Jeon, Yoon Seok, and Won, Chee Sun. 2000. Efficient use of local edge histogram descriptor. *Proceedings of the 2000 ACM Workshops on Multimedia.* ACM, pp. 51–54, Los Angeles, California.

Roweis, Sam T and Saul, Lawrence K. 2000. Nonlinear dimensionality reduction by locally linear embedding. *Science* (American Association for the Advancement of Science) 290: 2323–2326.

Simonyan, Karen and Zisserman, Andrew. 2014. *Very deep convolutional networks for large-scale image recognition.* arXiv preprint arXiv:1409.1556.

Soekhoe, Deepak, van der Putten, Peter, and Plaat, Aske. 2016. *On the Impact of Data Set Size in Transfer Learning Using Deep Neural Networks.* Leiden, the Netherlands: Leiden University.

Stein, Elias and Shakarchi, Rami. 2003. *Fourier Analysis: An Introduction.* Princeton, NJ: Princeton University Press.

Szegedy, Christian, Liu, Wei, Jia, Yangqing et al. 2015. Going deeper with convolutions. *Proceedings of the IEEE, Conference on Computer Vision and Pattern Recognition.* IEEE, pp. 1–9, Boston, Massachusetts.

Tenenbaum, Joshua B, De Silva, Vin, and Langford, John C. 2000. A global geometric framework for nonlinear dimensionality reduction. *Science* (American Association for the Advancement of Science) 290: 2319–2323.

Thorpe, Simon, Fize, Denis, Marlot, Catherine et al. 1996. Speed of processing in the human visual system. *Nature* 381: 520–522.

Tuck, Michael. 2010. Gestalt Principles Applied in Design. Retrieved Nov. 30, 2015.

Venkatesan, Ragav, Parag, Chandakkar, Baoxin, Li, et al. 2012. Classification of diabetic retinopathy images using multi-class multiple-instance learning based on color correlogram features. *Engineering in Medicine and Biology Society (EMBC), 2012 Annual International Conference of the IEEE.* IEEE, pp. 1462–1465, San Diego, California.

Von Helmholtz, Hermann. 1867. *Hanbuch Der Physiologischen Optik.* Scholar's Choice. Hamburg: Voss.

Vondrick, Carl, Khosla, Aditya, Malisiewicz, Tomasz et al. 2013. Hoggles: Visualizing object detection features. *Proceedings of the IEEE International Conference on Computer Vision*, pp. 1–8, Sydney, Australia.

Watamaniuk, Scott NJ and Duchon, Andrew. 1992. The human visual system averages speed information. *Vision Research* (Elsevier) 32: 931–941.

Zhang, Qiang and Li, Baoxin. 2010. Discriminative K-SVD for dictionary learning in face recognition. *IEEE Conference on Computer Vision and Pattern Recognition (CVPR).* IEEE, pp. 2691–2698, San Fransisco, California.

# 2
# LEARNING AS A
# REGRESSION PROBLEM

In Chapter 1, we saw some basic representations of the images and a few techniques of inference (mostly classification) based on those representations. It is evident from the discussion therein that machine learning has been a key to many computer vision tasks. Many of the machine learning tasks in computer vision may eventually be viewed as a general regression problem where the goal is to figure out a mapping from some input data to some output data. In the case of object/image classification, we are effectively finding a regression model from some feature vectors to some labels. A regression model may even be used to depict the mapping from some basic image representations, for instance, the raw pixel values, to some other abstract representations such as a compact real-valued feature vector. While Chapter 1 serves to only illustrate these particular examples, this chapter conducts a thorough and formal exploration of regression as a mathematical tool.

To facilitate the discussion of the key ideas without being burdened by the complexity of often high-dimensional image features, we pull back a little and create a more primitive feature space. We first present linear regression models for this illustrative feature space. Analytical solutions are developed, including ideas of introducing regularization to the basic formulation. Extension to the basic linear models is then discussed. The latter part of this chapter is devoted to iterative approaches to finding a solution, which is useful for real problems where finding analytical solutions is too difficult or not possible at all.

## Supervised Learning

The visual features we studied in Chapter 1, though intuitive, are a bit too sophisticated for a basic understanding of some of the fundamentals of machine learning. Let us create a more relatable and lower dimensional dataset for a more intuitive study of some fundamentals of machine learning. Assume that you are a human resource manager at a top data science firm and that you are planning to make strategic human resource expansions in your department. While interviewing candidates, you would like to know antecedently how much that candidate's pay scale is likely to be. In today's market where data scientists are in strong demand, most candidates have a free-market value they are predisposed to expect. As a data scientist yourself, you could model a future candidate's potential compensation and use this knowledge to negotiate during an interview.

A straightforward way to approach this problem is to use the compensation of all those who are already employed by your firm in estimating a future employee's pay. Say your company has $n+m$ employees. If you assemble a dataset of your current employees, you may come up with Equation 2.1 (considering for now only the first $n$ employees as the training data):

$$D = \begin{bmatrix} x_1 & y_1 \\ x_2 & y_2 \\ \vdots & \vdots \\ x_n & y_n \end{bmatrix} \tag{2.1}$$

where $x_i \in \mathbb{R}^d$ is a $d$-dimensional sample with each dimension corresponding to a particular value of an attribute of the employee and $y_i \in \mathbb{R}^1$ is the salary of the employee. An *attribute* of the employee could be his or her years of experience in the field, or his or her rank at stackoverflow.com, or the compensation of his or her previous position, and so forth. The rest of the $m$ samples that also have similar form are reserved for later use. $x_i$'s are often interchangeably referred to as *features, representations, samples, data, predictors, covariates, properties,* or *attributes.* They are called representations because $x_i$ represents the person. The representational space or the feature space, which in this case is $\mathbb{R}^d$, is a vector space that is closed with all possible candidates, each represented as a vector in this space. Each sample is a vector or a point in such a space. Similar to the image features that we discussed in Chapter 1, the features here describe the samples using attributes that relate to what we are trying to learn.

We use the terms *variates, targets,* or *labels* when referring to each $y_i$. These are what we are trying to learn. To *learn* is to establish a mapping between the features and the labels. To model the compensation of the employees, consider that $x_i \in \mathbb{R}^1$, a one-dimensional feature, perhaps the number of years of experience a candidate has in the field. The data might look something like that shown in Figure 2.1.

*Supervised learning* is the process of establishing a relationship or mapping through a model between $(x, y)$ using $(x_i, y_i) \in D \; \forall \; i \in [1, n]$ such that given a new sample $x_{n+j} \notin D$ and $j < m$, the model estimates $y_{n+j}$. In other words, we want to learn a model using a part of the dataset that we collected such that given a sample from the other part, we should be able to predict the associated label.

We call the dataset $D$ the *training dataset* and the set of samples $x_i$, $i \in (n, n+m]$ the *generalization dataset*. The generalization dataset is typically the real world. If we have the knowledge of the actual targets of the samples in the generalization set, we call it the testing set, as we can use it to evaluate the quality of our model before deploying it in the real world. A *model* is a functional transformation with a well-defined architecture that maps samples from the space of the features

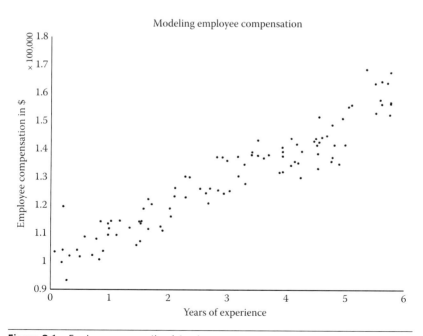

**Figure 2.1**   Employee compensation dataset.

to the space of the labels. Models are of the form $\hat{y} = g(X, w)$ where $w$ are the parameters or weights that transform $X$ to $\hat{y}$. Notationally, $X$ refers to rows of data vectors (columns) in matrix form and $\hat{y}$ refers to a list of predictions, one for each row in $X$. We use these notational systems, with slight abuse, to represent matrices and vectors. Note that we use $\hat{y}$ instead of $y$ here to distinguish between the predicted outputs and the ground truth labels from the dataset. In short, a model is a functional form that was predetermined that depends on some to be determined parameters whose values are to be learned using the data. Sometimes, this functional form is also referred to as a hypothesis.

The rationale for training from and testing on datasets is that we cannot, pragmatically, train from and test on the real world dynamically. The most important criterion of training and testing datasets is that they must be static, but as similar to the real world as possible. The formal way to define such a similarity is by using distributions. Suppose that the real world obeys a probability distribution for the data we observe. In our case, the salaries from the list of potential candidates requesting a particular salary and the list of extant candidates earning a salary follow some probability distribution. We assume that the distribution of the data we possess is similar if not the same. This is easy to realize in this example because our dataset itself is *sampled* from such a real world. A dataset that is sufficiently densely sampled will mimic the real world sufficiently accurately with few surprises. If we have infinite samples in a dataset, we can guarantee that the dataset mimics the real world. The more samples we draw supervision from, the more reliable our dataset, and ergo, the more reliable the models we learn from it (Button et al., 2013). It is to be noted, however, that in machine learning, we are only trying to learn a fit to the dataset with the hope and understanding that such a fit is the best we could do to fit the real world. In this perspective, supervised machine learning models mostly the distribution of the real world by obtaining an approximation using the given training dataset that we have sampled.

### Linear Models

Let us now make an assumption that simplifies the problem: Let us posit that the experience of the candidates and their compensation are linearly related. What this means is that the relationship between the candidates' experience and the salaries is captured by a straight line. With this assumption, we have converted this problem into a linear regression problem. Linear regression and its variants are some of the most commonly used models in data analysis (Neter et al., 1996; Seber and Lee, 2012; Galton, 1894). With the assumption that our data and the labels are linearly dependent, let us create $g$ as a linear model: If $x \in \mathbb{R}^1$ then

$$\hat{y} = g(x, \boldsymbol{w}) = w_1 x + b \tag{2.2}$$

If $\boldsymbol{x} \in \mathbb{R}^d$ then

$$\hat{y} = \sum_{i=1}^{d} w_i x^i + b \tag{2.3}$$

$w_i$ are the parameters of the linear model $g$ and they represent the norm of the underlying hyperplane fitting the data. Customarily, $b$ is considered a part of $\boldsymbol{w}$, such that $b = w_0$ and $x^0 = 1$. With these conditions, we can write Equation 2.3 in matrix form as

$$\hat{y} = \boldsymbol{w}^T \boldsymbol{x} \tag{2.4}$$

It appears from Figure 2.1 that the base salary with 0 years of experience was about \$100,000 and for every year there seems to be an additional \$10,000. With this knowledge, we can now make a model with $w_1 = 10,000$ and $w_0 = 100,000$. Figure 2.2 shows a linear model that fits the data with these parameters.

It is one thing to look at a line or curve in a two-dimensional dataset and estimate the model parameters (simply the slope of the line and its intercept with vertical axis in this case), but it is another task entirely to look at a multidimensional dataset and estimate the general weight vector of Equation 2.4. To be able to fit linear models to larger and high-dimensional datasets, we would like to have a formal way to estimate these parameters. The process of estimating these parameters such that we get a good fit to a dataset is called *training*.

Initially, given just the dataset and armed with the knowledge that we are dealing with a linear model, we do not know what the model

**Figure 2.2**   Linear fit for the employee compensation dataset.

parameters ought to be. In simplistic cases, one might be able to write down a system of linear equations and solve for a unique solution. However, for most real datasets, a solution is typically obtained by an iterative process: We start with some initial guess for a solution and then iteratively improve upon the solution. We could initialize the model with some random numbers. Unless we are extremely lucky, the randomly created model is sure to produce some, if not a lot of, wrong predictions for the target. We can then adjust the parameters of the model after observing the wrong predictions such that we get better and more informed predictions the next time around. We should repeat this process until such a time that no matter how we tune, we do not improve on our predictions on the targets. This process is illustrated in Figure 2.3. Once we have the model trained, we can continue on to the prediction of the labels (Figure 2.4).

   We are still not much further along if we have to tweak the parameters randomly until we make good predictions. Ideally, we would like to make a decision about which direction to tune our parameters so that every time we make an adjustment, we improve upon our previous predictions. While we elaborate more on this *tuning–based* system later in this chapter, we explore some straightforward ways of estimating the model parameters in the next section.

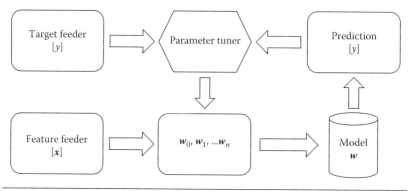

**Figure 2.3**  An illustration of training a model.

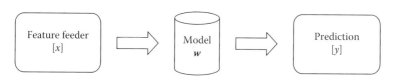

**Figure 2.4**  An illustration of predicting a label given a model.

## Least Squares

Let us look at the training protocol in Figure 2.3. Since we began with randomly initialized weights $w$, our predictions $\hat{y}$ would also have been random. Let us create a measure of how wrong our initial blind predictions were. An error function $e_i$ of the form

$$e_i(w) = \left\| \hat{y}_i - y_i \right\|_2 \tag{2.5}$$

will tell us how far away our prediction $\hat{y}_i$ is from the actual value $y_i$, $\forall i \in [0, n]$ in the Euclidean sense. This implies that given some set of parameters $w$, the error as described in Equation 2.5 gives us a measure of *how wrong we were* with our predictions. The functional form of the error ensures the error is positive, because we take the square of the difference.

In Figure 2.5, the solid line is a *better* model than any of the broken lines. The *better* quality can be described in the following manner: Equation 2.6 describes the error for each point as the distance from the point to the line. This is also the error that the model is off by in its prediction of that point. This is a Euclidean measure of *how far* away the model's predictions are from the true samples. The solid line is a

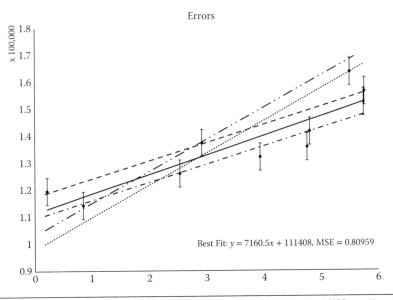

**Figure 2.5**   It can be seen that the solid line produces the best fit with the least MSE even though there are other lines that pass through more than one data sample.

better model than the broken line because the solid line has a smaller accumulated error than the broken lines.

Cumulatively, the solid line makes better predictions than the broken lines even though the solid line passed through none of the samples directly. The broken lines may pass through at least one sample, but overall they still have more accumulated error. The cumulative error on the entire training dataset can be formally defined as

$$e(\boldsymbol{w}) = \sum_{i=1}^{n} || y_i - \hat{y}_i ||^2 \qquad (2.6)$$

The cumulative error is often referred to as *objective, error, loss, cost,* or *energy* interchangeably. Our learning goal is now to find that model that gives the least possible error $e$, that is, the *least squares* solution (Lawson and Hanson, 1995). More formally, we want the parameters $\boldsymbol{w} = \ddot{\boldsymbol{w}}$ such that

$$\ddot{\boldsymbol{w}} = \operatorname{argmin}_{w} e(\boldsymbol{w}) \qquad (2.7)$$

Following our earlier matrix notation, Equation 2.7 can be written as

$$e(w) = (y - w^T X)^T (y - w^T X) \qquad (2.8)$$

Fortunately for us, there exists an analytical solution for Equation 2.7. The derivation of this analytical solution takes a little linear algebra and is left to the reader. The analytical solution for the optimal parameter set is

$$\ddot{w} = (X^T X)^{-1} X^T y \qquad (2.9)$$

where $\ddot{w}$ are the best set of weights that minimizes Equation 2.9. This solution, of course, assumes that the matrix inversion in Equation 2.9 is possible.

Although for this simple linear model an analytical solution does exist, we find that for more complex problem structures we have to rely on some optimization procedures that are described in the later sections. For the moment we shall stick with this analytical solution and study linear regression in these settings. Once we have the model solution, we can make predictions using this model for any sample $x_{n+j}$ as $y_{n+j} = \ddot{w}^T x_{n+j}$. The task of making these predictions is called testing and is shown in Figure 2.4.

### Maximum-Likelihood Interpretation

Let us rethink the idea of distances. Consider a 1D Gaussian distribution at each point on the line with a variance $\sigma^2$. If the model is well-trained (i.e., producing a prediction very close to the ground truth), the value of the Gaussian at a certain distance from the line describes the probability that a point shall be found at that distance. We can also observe that the farther away we move from the line, the less probable that we might find a sample. The Gaussian describes this probability of finding an actual sample as we go farther away from the model. Hence, solving least squares is the equivalent of maximizing the likelihood of the model given the training samples.

Figure 2.6 shows this interpretation. In this figure, the dotted lines that connect each point to the Gaussian give the probability of finding that point there. Note that none of the Gaussians go below the 0 plane. The farther away the sample is, the less probable it is that it would exist. In this interpretation, we can see that the learning task now is to create a line segment such that all the heights (dotted lines)

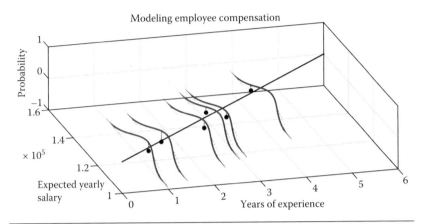

**Figure 2.6**   The probability of observing a sample is modeled by a Gaussian function of the distance to the true model line.

that show the probability that the points are on the line are cumulatively maximized. In other words, we want to choose that model that gives us the *maximum likelihood* of finding the given samples.

Let us formalize this viewpoint further. Consider Equation 2.10:

$$P(X, w) = \mathcal{N}(w^T X, \sigma^2) \tag{2.10}$$

Equation 2.10 is the probabilistic model of linear regression as shown in Figure 2.6. The mean of the Gaussian is simply the points on the line $w^T X$, which is our model. Assuming that all the samples are identical and independently distributed (i.i.d.), we can rewrite Equation 2.10 as

$$P(X, w) = \prod_{i=1}^{n} \mathcal{N}(w^T x_i, \sigma^2) \tag{2.11}$$

Let us now create a log-likelihood: the log of the likelihood of the existence of a data point given the line itself is.

$$\log P(X|w) = \sum_{i=1}^{n} \log P(y_i \mid x_i, w). \tag{2.12}$$

Maximizing a log-likelihood is an often-used function or tool in machine learning (Nadaraya, 1964). Likelihood is often the property that we seek or desire: in our case, the cumulative probability of finding samples closer to our model. The logarithm is a monotonous function, which implies that maximizing the likelihood is the same as maximizing the log of the likelihood. Instead of wanting to find a line that best fits the model, in maximum likelihood estimation (MLE), we want to find a model that best realizes the likelihood of the dataset existing in its configuration possible.

Since our model is a Gaussian, we have

$$l(\boldsymbol{w}) = \sum_{i=1}^{n} \log\left[ \left( \frac{1}{2\pi\sigma^2} \right)^2 \exp\left( -\frac{1}{2\sigma^2}(y_i - \boldsymbol{w}^T \boldsymbol{x}_i) \right) \right] \quad (2.13)$$

Fortunately for us, it turns out that maximizing Equation 2.13 is equivalent to minimizing Equation 2.14, which is exactly the same as minimizing Equation 2.7:

$$e = \sum_{i=1}^{n} (y_i - \boldsymbol{w}^T x_i)^2. \quad (2.14)$$

If prediction using the linear regressor is the evaluation of the line at the point, prediction using the MLE is the evaluation of the Gaussian at that point. The advantage of using MLE is that, along with a prediction, we also get a confidence of prediction (height of the Gaussian) for any guess we make.

### Extension to Nonlinear Models

Thus far, we have only seen cases where the models are a simple straight line. In this section, we deal with the case where the linearity assumption that we made might not hold true. Since we are comfortable with linear models, when a dataset is not well-modeled by a linear model, we attempt to convert it into one for which a linear model might just work. These and other *kernel* methods are among the most popular methods outside of neural networks in the field of computer vision (Shawe-Taylor and Cristianini, 2004; Ramsay and Silverman, 2006; Vapnik et al., 1997; Gunn, 1998).

Expanding a linear model to a nonlinear one may be done conceptually by simply replacing $w^T X$ by $w^T \Phi(X)$ where $\Phi$ is a nonlinear mapping of $X$. For instance, by performing the regression

$$\hat{y} = w_1 \sqrt[2]{x} + w_2 x + b \qquad (2.15)$$

we are effectively performing a linear regression with $\Phi(x) = \left[ \sqrt[2]{x}, x \right]$. Note that the relation between $x$ and $y$ is no longer linear, although we may still perform linear regression through the introduction of $\Phi(x)$. This transformation is called the basis function expansion.

With the use of basis functions, we are projecting the data onto a new space upon which we expect linear models to make better predictions. Once we move on to neural networks in later chapters, we study automated ways of learning the functional form of such a projection $\Phi$. But at this moment, we consider making predictions using explicit higher order models. The capability to perform regression on arbitrary model complexities comes with its own basket of problems.

The first of those problems is the phenomenon that we notice in Figure 2.7. The problem arises from the fact that the higher the order of the model, the more complex our fit is and the more points through which we can pass. Although it sounds like a good idea to have more points to pass through, this is not often the best minimizer of the cumulative error as we have already noticed in Figure 2.5.

All these curves that twist and turn means that for some test cases, we may predict wildly different values. Consider, for instance, the data point marked ★ in Figure 2.7. This employee's compensation would have been predicted *right on the money* if we used a linear model rather than a model that is a sixth-order polynomial. In the sixth-order polynomial fit, we have grossly underappreciated our interviewee, who would be wise to reject our offer. This problem is called *overfitting*. Overfitting is one of the most common causes of deployment failures of many learning systems. Overfitting will produce tremendous performance on predicting results from our training set itself but will grossly underperform in our testing set. Overfitting is also easy to spot. We are with all probability overfitting if we make extremely

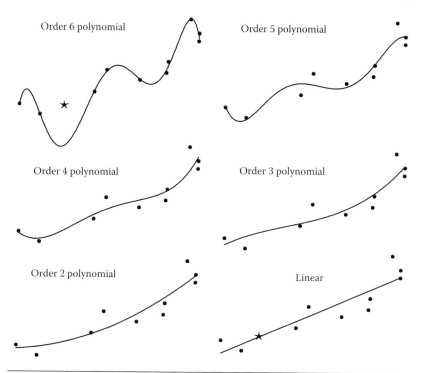

**Figure 2.7**   Regression with polynomial basis functions of various orders. Note that the models shown here are on the original vector space and therefore the nonlinearity. These are learned on a transformed feature space, where the positions of the features would be different for different orders while the models would all have been linear. An illustration on this transformed space is not useful to study though as we cannot make meaningful comparisons. Therefore, the linear model that we learned on the transformed space is projected onto the original space using the inverse mapping that gives us these nonlinear curved models.

accurate predictions for a few samples and extremely poor predictions for others. We are almost certainly overfitting if we are producing an extraordinary performance on our training set and an indifferent performance on our testing set. One simple way to avoid overfitting is by regularization.

### Regularization

For a curve that has to twist and turn a lot, the coefficients (weights) of that curve have to also wildly swing between a large negative value and a large positive value. To avoid such high twists in curves, and therefore prevent overfitting, we can introduce additional penalties to the loss function. There are different ways of regularizing the learning with additional penalty terms in the error function. Let us first explore the idea of a *penalty*. Whenever there is a function that we are trying to minimize (or maximize), we can make it obey certain self-imposed constraints and traits that we want by adding a weighted measure of what we want to constrain onto the loss function. Let us now use this intuition to constrain the weights from not taking larger values. Let us rewrite Equation 2.15 as

$$e = \| y - w^T x \| \tag{2.16}$$

the error function that we are trying to minimize. To ensure that the weights do not explode, we could apply a *regularizer* that penalizes Equation 2.17 for having larger weight values:

$$e = \| y - w^T x \| + \alpha_2 w^T w \tag{2.17}$$

$w^T w$ is the length of the weight vector. The ultimate objective of learning is to reduce $e$. Whatever we add on to $e$ is also something that we are consciously trying to minimize. If what we add is either a constant or is something that is not related to the prediction of labels given the dataset, it will not affect the way we learn. In Equation 2.17, for instance, we have added a function of the weights itself to the loss. This enables us to keep a check on the values of the weights from exploding.

In this case, the second term should not contain $w_0$, as we do not worry about the intercept of the function being large. $\alpha_2$ is a factor that balances the numerical values of the loss and the penalty so that one does not get lost by the magnitude of the other. Having a large $\alpha_2$ would force the weights to be small but not care about the errors much; having a small $\alpha_2$ will do the opposite. $\alpha_2$ is essentially the trade-off for what we want the learner to focus on. This penalty ensures that the weights remain small while the error is also minimized. The minimization function is a trade-off between wanting a better fit (first term) and wanting a smaller set of weights (second term).

It turns out that even for this problem setup, we still have an analytic solution:

$$w = (X^T X + \alpha_2 I_d)^{-1} X^T y \tag{2.18}$$

where $I_d$ is the identity matrix of the required size. This solution also has the nice property that the matrix inverse term is much more stable during inversion than the one from Equation 2.9 and this more often than not leads to a better solution than a simple linear regression.

This process is called *regularization* because it regulates the function from undesirable weight explosions. This particular type of regularization is called $L_2$ regularization as we penalize the loss function with the $L_2$ norm of the weight vectors. This type of regression is also often referred to as ridge regression (Hoerl and Kennard, 1970; Le Cessie and Van Houwelingen, 1992; Marquardt and Snee, 1975).

We could also, for instance, consider reducing the complexity of the model itself as another form of regularization. Consider that we are trying to fit a sixth-order polynomial as before, but if we could *afford* it, we would much rather prefer a smaller model. We could enforce this by penalizing the number of nonzero weights instead of the magnitude of the weights as in the case of $L_2$ regularization. The more nonzero coefficients we have, the higher the order of our model. We can accomplish this by using the $L_1$ norm instead of the $L_2$ norm as follows:

$$e = \| y - w^T x \| + \alpha_1 | w |_1 \qquad (2.19)$$

This regularizer is often called the $L_1$ regularizer. This ensures that we have a *sparse* set of weights. To minimize Equation 2.19, we need as low a second term as possible. The only way to lower the second term is by having the weights go to zero. By making a weight go to zero, we are enforcing sparsity among the weights. Having a sparse set of weights implies that if all the weights were assembled as vectors, most of the weights will be absolute zeros. This is a very aggressive regularizer.

By having a sparse set of weights, we are also choosing which attributes are needed and which are not needed for our curve-fitting. By applying sparsity therefore, we are also performing *feature selection*. An $L_1$ regularizer helps us in being able to always start off with an extremely complex model and then simply regularize until we can throw away as many features as we can afford to and still acquire good generalization performance.

Why does regularization work? Shouldn't it be the case that having more complex and more sophisticated solutions to a problem must be helpful? The answer to that question is succinctly answered by the *lex parsimoniae* philosophy of Ockham's razor (Jefferys and Burger, 1992). It claims that *the simplest of solutions is often the most accurate.* More precisely, the statement of Ockham's razor is *among competing hypotheses, the one with the fewest assumptions should be selected.* Regularization extrinsically discourages very complicated models and eliminates as many variables as possible in our predictions. Extreme explanations of the world even if they fit the data we observe are not always the correct explanations. Extreme explanations and complex models do not usually generalize well to future testing data. They may work on a few samples well, but this may just be because of the idiosyncrasies of those particular samples we trained on. We should choose to have a slightly less desirable fit for an entire dataset than a perfect fit for the idiosyncrasies of a few samples.

### Cross-Validation

We have far outgrown what started out as a tuner to tune coefficients of a linear model into a system in which we also need to choose a lot of other *hyperparameters* including model size, weights for regularizers, and type of regularizer. While the weights are often learned from dataset, other model parameters, such as regularizer weighting, which are not related directly to predictors, have to be chosen manually. In order to accurately choose these hyperparameters, we can make use of our original training dataset itself in a smart manner.

Let us divide our original training dataset into two parts: a validation dataset and a reduced training dataset. Suppose we are picking one of these hyperparameters (say $\alpha_1$). We can fix a certain value for $\alpha_1$ and proceed to train on the reduced training set. We can evaluate the performance of our model on the validation set. The best parameter setting is the setting that provided the best performance on the validation set. Note that in this case, the validation set must be disjoint from the reduced training set and usually should remain consistent. The validation set is usually smaller than the training set and of the

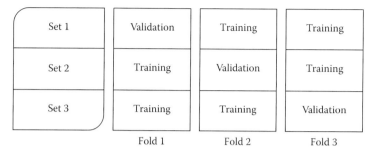

| | Fold 1 | Fold 2 | Fold 3 |
|---|---|---|---|
| Set 1 | Validation | Training | Training |
| Set 2 | Training | Validation | Training |
| Set 3 | Training | Training | Validation |

**Figure 2.8**  Cross-validation.

same distribution as the testing set. The validation set is also expected to have the same distribution as the training and testing sets.

We could also create many models with various parts of the training set playing the role of a validation set for each model. Our final prediction is now the average of all the predictions of all the models. These kinds of cross-validations are called $k$-fold cross-validations (Kohavi, 1995; Efron and Gong, 1983). The technique of cross-validation was particularly influential in the development of ridge regression in fixing the hyperparameter of the norm weights and later utilized on other linear methods and model selection in general (Golub et al., 1979; Shao, 1993; Kohavi, 1995).

Figure 2.8 shows a setup for a threefold cross-validation. Once we divide the dataset into three disjoint parts, we start off with the first set playing validation and the others the training; we can then rotate this policy until all parts of the dataset have played the role of validation once. This gives us $k = 3$ models and our final prediction is the average prediction of the three models thus created. An extreme version of this cross-validation is the leave-one-out cross-validation (LOOCV) (Zhang, 1993). If the training set has $n$ samples, we make $n-1$ samples as training data and the remaining sample is the validation data.

**Gradient Descent**

In the previous section, we studied the building of the linear regression model. In doing so, we focused on analytical solutions and consciously left undiscussed an alternate method of arriving at the solution: how to iteratively tune the weights so as to get the least error. The mathematical processes in such training that involves twiddling the weights in just the right way to achieve the goal of minimizing the error dependent on the weights being twiddled with are often gradient-based optimization techniques.

We discussed an analytical solution for the linear regression problem earlier, but before we wrote down the analytical solution and without it, we were forced to randomly twiddle with the weights until we achieved some satisfactory performance on some test dataset (Boyd and Vandenberghe, 2004). The analytical solution may not be feasible for many instances of the linear regression problem. To be able to perform the inverse or the pseudoinverse, we require a large amount of memory and computation. We also require a guarantee that the inversion is stable and nonsingular. Also, as we shall see in future chapters, the analytical solution holds only for the linear regression model. If we were to do anything sophisticated like adding a few more layers of nonlinear mapping, we are out of luck.

Data as described in Chapter 1 are static. While this was true for the case of employee compensation in a company, it is rarely true in the real world. Data from the real world are typically dynamic and are often streaming. Even if the data were not streaming by nature, due to insufficiency of digital computing memory, data can at best be processed in batches. To use the analytical solution, we cannot have streaming or batch data and need all of it at once. Also, for our previous analytical solution, if the data changes we need to solve anew for the weights, since our solution might no longer remain optimal.

With the problems discussed above, in this section, we shall seek a practical solution for obtaining $\ddot{w}$. In doing so, we will also find some unique and serendipitous capabilities of the parameters at which we arrive.

Let us look back at the error function that we forged in the previous section:

$$e(w) = (y - w^T X)^T (y - w^T X) \qquad (2.20)$$

This error that we created is a *bowl*-shaped function in the parameter space. Let us visualize this and have a closer look. Figure 2.9 shows such a visualization for 2D data. In the figure, the axes $w_0$ and $w_1$ represent the parameters of the model. Any point on this plane is a particular configuration of the machine itself. A configuration of the machine is a state of existence of the regressor having already assigned

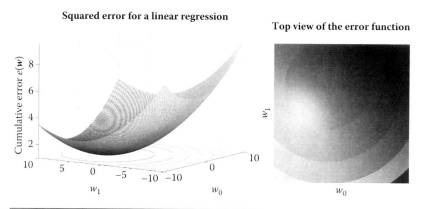

**Figure 2.9**   The *bowl* shape of the error function. On the left is a 3D visualization with its contours projected on the floor. The right is a top view of the same function. In both figures, the circumference of the contour is a locus of points with the same loss value. Moving along the circumference does not affect the loss; moving in a direction that is perpendicular to the tangent, inward is the direction of the steepest descent.

a particular set of values for the weights. The value of the weights is itself a configuration.

Within the space of $w_0$ and $w_1$ there exists one configuration of the machine that gives the least error, which in this case is $\ddot{w} = [\ddot{w}_0, \ddot{w}_1] = [3, 3]$. The solution is $[3, 3]$ because that is the configuration at which the machine reaches the lowest possible error value. The machine is a two-parameter linear regressor. It is to be noted, though, that we hypothesized that the features and the output we are trying to learn are linearly dependent. As long as this linearity hypothesis holds, the solution is a configuration somewhere in this space. The solution found may not be perfect over the training or the testing datasets, but there is at least one solution that is optimal. The term *hypothesis* is often used in machine learning and computer vision literature in reference to the type or the functional form of the model. Given a different hypothesis, say, a quadratic feature thrown in through a basis function expansion, we might or might not be able to get a better error than the one we got from our linear hypothesis.

Suppose that the model we are making is a two-parameter linear system with parameters $w_0$ and $w_1$. In Figure 2.9, the optimal solution we seek is $\ddot{w} = [\ddot{w}_0, \ddot{w}_1] = [3, 3]$. An optimal solution is one that, given the hypothesis, produces the least error among all possible configurations of the hypothesis.

In the training protocol we discussed in the previous sections prior to settling on the analytical solution, we started with a random set of initial parameters and twiddled with them until such a time we reached the optimal solution. In the subsequent sections, we will establish a formal procedure and specific techniques for doing the weight-seeking iteratively.

The gradient of a function with respect to one of its components is the rate at which the function changes with a small change of that component. It indicates the direction of greatest change. The quantized version of the gradient that we should suppose contextually here is the Newtonian difference. For a function $e(w)$, the gradient can be defined by its first principle, $e'(w) = \dfrac{e(w + \varepsilon) - e(w)}{\varepsilon}$, where $\varepsilon$ is a small change in the parameters $w$. Note that this gradient is estimated for some location $w$ by perturbing it slightly by $\varepsilon$ and observing the change.

Typically, we do not have to depend upon this first principle perturbation action to obtain a gradient. Most regression and neural architectures have a closed form for their gradients. The gradient of Equation 2.8, for instance, is

$$e'(w) = -2X^T Y + 2X^T X w \qquad (2.21)$$

Note that this gradient is a component-wise vector. Each component of this gradient vector literally measures how much the error is affected by a small change in that weight component.

Now suppose the function were one of our error functions: a bowl such as the one shown in Figure 2.9. Any point in this surface that is not at the bottom of the valley will have a gradient that is pointing outward and away from the bottom of the valley. That is the direction in which the function is rising maximally. Since our aim is to minimize the error, we do not want to change the weights in this direction. In fact, we want to change the weights in the exact opposite direction. Note that at the lowest point of the error function, the desirable configuration that we seek, the gradient is always zero. If we find ourselves in a weight configuration where no matter which direction we look the gradient is pointing outwards and rising, we have reached the minimum location.

In other words, gradient measures the curvature or the slope of the function at a point and the minimum is at the point of zero slope.

Consider this scenario now developing as follows: We start by initializing our machine with some random configuration. This random configuration corresponds to some point in the space of $w$'s. Unless we are extremely lucky and pick the configuration at the center of the bowl, we will always be able to do better. Once present at some location in the space of $w$'s, we look around the space and find the direction of the steepest descent, which is the direction opposite to that of the gradient at that location.

We take a step in that direction by subtracting, from the weights, a scaled version of the gradient due to that weight component. We look around again and measure the gradient at the point. Unless we are at a point where the curvature is no longer decreasing, we keep repeating this process. If we find ourselves at a point with the curvature no longer decreasing in any direction, we have arrived at the valley of the bowl.

Thus, we have developed an iterative *learning* algorithm: *the gradient descent*. More formally, if at some iteration $\tau$ we are at some location $w^\tau$, we can now update the weights as

$$w^{\tau+1} = w^\tau - \eta^\tau e'(w^\tau) \tag{2.22}$$

The true nature of the objective function $e$ is unknown. Had we known it, we could quite simply have picked the minimum value. Instead, given the data $(X, Y)$, we can only make an approximation to the error at some location $w^\tau$ at some iteration $\tau$. The gradient of the accumulated error given all the data samples is also therefore a close approximation to the gradient of the actual error function and not its actual gradient.

Naturally, the more data we have the better our approximations will be. If we have infinite data, we might be able to generate the true gradient. Alternatively, the less data we have the noisier our gradient approximations are. Figure 2.10 shows the typical path taken by a typical gradient descent algorithm. It is easy to note that the general direction of motion is pointed toward the center of the bowl, which is the point we seek.

Gradient descent

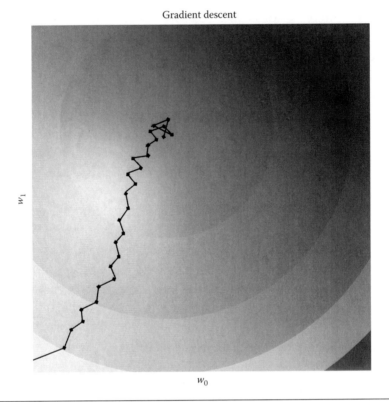

**Figure 2.10**   Path taken to the center by the gradient descent algorithm. The rings are contours of equal error and we are looking from atop the parameters plane into the error function. The black line is a step for each iteration and the squares are the configuration at each iteration.

The parameter $\eta$ is often referred to as the learning rate. The choice of $\eta$ plays a major role in the learning algorithm. If $\eta$ were too small, we traverse the space at a leisurely pace; if $\eta$ were too large, we take larger steps. Both strategies have their own pros and cons. Taking larger steps is suitable when we are certain about our approximation of the gradient. We do not want to take larger steps when we are not certain of the direction in which we are taking that step. The approximated gradients are noisy when we use data in smaller batches or if we do not have a large dataset. With a large $\eta$, we also run into the problem of bouncing around a local minimum without actually converging at it. As can be seen in Figure 2.10, if the minima is within the size of our step, we would overshoot around it. This creates the effect of ricocheting around the local minima without actually reaching it. If η were too small, we would take a lot of iterations to reach the mini-mum if we were able to reach it at all.

Ideally, we would want a dynamically changing η, hence we have $η^τ$ dependent on τ in Equation 2.22. A commonly used strategy for choosing η is to pick one that is sufficiently large at an early stage so that we can proceed faster, but as we go through a few iterations, we make it smaller so that we can converge. Usually, this reduction is chosen to be a function of the iteration itself. More often than not, $η^τ$ is piecewise linear or exponential in τ.

The second-order derivative or the Hessian of the error function, if it exists, is an indicator of the curvature of the error manifold. Using the inverse of the Hessian or the Taylor approximation of one is a popular way to avoid choosing the learning rate in this fashion. The curvier the space is, the larger a step we should take. The function gets flatter as we approach the minimum and is perfectly flat at the minima. Putting together, these techniques were popularized for machine learning, and gradient descent has been among the most preferred algorithms for learning neural networks (Rumelhart et al., 1986).

## Geometry of Regularization

In the previous section, we noticed that allowing the weights to be arbitrarily large introduces structure in regression that is undesirable and leads to overfitting. To avoid such problems, we briefly discussed the concept of regularization. In this section, we shall revisit regular-ization and its geometry in the context of optimization using gradient descent.

Consider $L_2$ regularization first. $L_2$ regularization is performed by adding $α_2 w^T w$ to the right-hand side of Equation 2.8. By per-forming $L_2$ regularization, we are limiting the weights to be con-strained within a circle with its center at the origin. $w^T w$ is a circle (hypersphere). This additional term in the loss function is going to increase if the radius of the circle grows larger. In a feature space (two-dimensional as shown in Figure 2.11), this would make the loss function larger if the weights were outside of the circle, whose radius is what is being carefully limited. In a high-dimensional feature space this extends to a hypersphere with the same philosophy. This limits

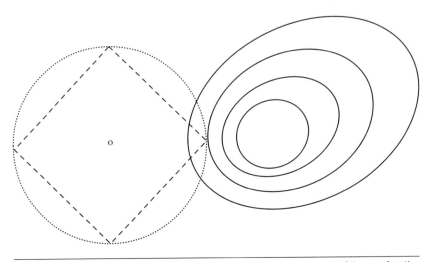

**Figure 2.11**    Geometry of regularization. The concentric ellipses are contours of the error function from atop. Each contour is the locus of points that has the same error value. The square with dashed lines is the $L_1$ constraint and the circle in dotted lines is the $L_2$ constraint. The axes in this figure are the parameters $w_0$ and $w_1$. The point 'o' is the origin of the space.

the weights from growing larger overall. The idea of this constraint is visualized in Figure 2.11.

The optimal solution, given this constraint, is always going to be on the circumference of the circle. Any point inside it is going to have a higher error since any point in the interior of the circle is moving away from the center of the error function's valley. There is a balancing act between two needs in play here. The regularizer is going to push the weights to be as close to the origin as possible, and the loss function will push the weights to be as close to the error's valley as possible. Therefore, the optimal solution has to be at the extreme edge of the shape of the regularizer, which in this case is the circumference of the circle.

This property is illustrated in Figure 2.11, where clearly the only best solution given the circle constraint is the point where its circumference meets one of the error contours. There is simply one point at which the error function meets the regularizer in this case. No error contour that has a lower error than that contact point would ever be inside the circle. This limits our search space to the points on the circumference of the circle. $L_2$ regularization imposes that cross terms in a polynomial combination of weights do not exist with a high probability. Since in our search for the optimal parameters we constrain ourselves to the circumference of a hypersphere, we are reducing the sizes of the models we are considering.

$L_1$ regularization, on the other hand, as we saw in the previous sections, is a much stronger imposer of sparsity of weights. We noticed that the $L_1$ regularizer imposes a much higher penalty for even having nonzero weights than $L_2$, which simply limits the weights from exceeding a certain value. The geometry of the $L_1$ regularizer is a square with its center of gravity at the origin and is also shown in Figure 2.11. Although we do not prove it mathematically here, the lasso ($L_1$) has a much higher probability that the error function and the constraint square interact most often at the vertices of the square. The vertices of the square always lie on the axes, implying that at least some, if not most, of the weights will be forced to go to zero. This is because the vertices lie on the axes. On the axes, the weight that corresponds to the axes is zero, hence we are able to impose sparsity. These are often called as weight decay techniques.

### Nonconvex Error Surfaces

So far, we have studied gradient descent in the context of a convex bowl error surface. A convex bowl error surface implies that we have one optimal solution that exists globally in the feature space. In general, the error surfaces of complex systems like neural networks have more than one peak and valley. A typical error surface with several peaks and valleys is shown in Figure 2.12.

We can still continue to use our old gradient descent technique to find a *good* bottom of the valley point. We should be aware though

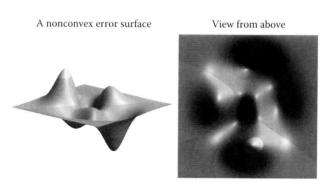

A nonconvex error surface          View from above

**Figure 2.12**    A more complex, traditional, and typical peaks and valleys nonconvex error function.

that there are many such valleys that are available for us to reach to the bottom of. We could end up in any one of them. We might want to run gradient descent a few times and cross-validate to find a good enough valley. Where we end up depends on our initialization as can be observed in Figure 2.12.

Since in these nonconvex surfaces initial conditions matter a lot, several recent methods have been proposed to smartly initialize large systems. We shall see some of them in upcoming chapters. Recent studies have also shown anticlimactically that in large and deep neural and convolutional neural networks (CNNs), often all valleys are of similar depth. This means that using any one of those valleys as a solution is the same as using any other (Choromanska et al., 2015).

### Stochastic, Batch, and Online Gradient Descent

So far we have used all the samples in our data $(X, y)$ in Equation 2.21 to calculate the error gradient approximation and therefore to update the weights in Equation 2.22. As we have seen before, it is not always feasible that we will have the entire dataset on demand. We should be satisfied by processing them in batches. One simple and possibly the most extreme version of this technique is what is referred to as the stochastic gradient descent. Recollect that our training dataset had $n$ samples. We can make our gradient estimate sample by sample. Feed one sample through the neuron, check its output, measure its error, and produce the gradient approximation. Let us rewrite our learning equation for this setup:

$$w^{n\tau+i,} = w^{n\tau+i-1} - \eta^{n\tau+i-1} e'(w^{n\tau+i-1}) \qquad (2.23)$$

where

$$e'(w^{n\tau+i-1}) = -2x_i^T y_i + 2x_i^T x_i w^{n\tau+i-1} \qquad (2.24)$$

Here, the $n\tau$ iterations count to what is often referred to as an epoch. One epoch is all the set of iterations that goes through the entire dataset once. In our previous setting (Equation 2.22), each iteration is an epoch. We call this old process batch gradient descent because we process them all in one batch.

The error so calculated by Equation 2.24 is going to be extremely noisy and extremely unreliable. To compensate for this effect, we reduce the learning rate drastically and move much slower through the space. This is preferable and still works fast enough because we take one step per sample rather than one step per epoch. While in batch gradient descent we make one confident large step per epoch, in stochastic gradient descent we make $n$ small and dubious steps.

There exists an obvious middle ground: Divide the dataset into mini-batches and estimate a better-than-stochastic but still diffident gradient and learn by a process similar to the one described above. In modern-day CNNs and deep learning systems, it is preferable to work in mini-batches due to architecture, memory, and system design factors. With mini-batches, we can have a reasonably high learning rate and still take a lot of steps. Modern-day GPUs are designed in such a way that for CNNs, mini-batches are more productive than any other technique.

### Alternative Update Rules Using Adaptive Learning Rates

It was briefly mentioned earlier that one way to avoid picking a learning rate and quicken moving through the error manifold was to use the inverse of the Hessian as the learning rate. This is often referred to as Newton's method. Consider the following update rule:

$$w_i^{\tau+1} = w_i^{\tau} - H_i^{-1} e'(w^{\tau}) \qquad (2.25)$$

Here $H_i^{-1}$ stands for the Hessian of the $i^{\text{th}}$ weight vector. Using the Hessian is much faster and, although we do not show it explicitly here, requires only one iteration if we were in a convex bowl error function. Despite its nice properties, Newton's method is often not preferred because the Hessian is a large matrix, oftentimes larger than what most memory systems allow. The Hessian has exponentially more terms than the number of parameters and is not suitable for deep learning systems where there are a lot of parameters.

Some alternative second-order update methods do exist with linear-in-parameter-count memory requirements. One such popular method is Adagrad and its update rule is shown in Equation 2.26:

$$w_i^{\tau+1} = w_i^{\tau} - \frac{\eta^{\tau}}{\sqrt[2]{\sum_{t=1}^{\tau} e'(w_i^t)^2 + \epsilon}} e'(w_i^{\tau}) \qquad (2.26)$$

One can view the entire term $\dfrac{\eta^\tau}{\sqrt[2]{\sum_{t=1}^{\tau} e'(w_i^t)^2 + \epsilon}}$ as a learning rate

that is dependent on previous gradients. Using the summation on $t$, we are collecting and accumulating the second-order gradients over previous iterations. This accumulated gradient contains knowledge about the average direction in which the gradient has been moving up and until the current iteration. This way Adagrad reduces the burden on the choice of the learning rate (Duchi et al., 2011). $\epsilon$ is added largely as a fudge factor to avoid division by zero in case the gradients are flawed or zero. The step size is largely controlled by the size of the gradients produced in the previous iterations. This method also essentially has an independent learning rate per parameter that helps in faster convergence. The size of the step we take in each independent dimension is not the same but is determined by the gradient accumulated in that dimension from the previous iterations. Note that the denominator is a vector, with one value for each dimension of the gradient.

Root-mean-squared propagation (RMSPROP) is an adaptation on Adagrad that maintains a moving average of the gradients (Hinton, n.d.). The RMSPROP update rule is

$$w_i^{\tau+1} = w_i^\tau - \frac{\eta^\tau}{\sqrt[2]{m_\tau^i + \epsilon}} e'(w_i^\tau) \tag{2.27}$$

where

$$m_\tau^i = \rho m_{\tau-1}^i + (1-\rho)e'(w_i^\tau) \tag{2.28}$$

and $\rho$ must be between 0 and 1. This works in a fashion that is similar to Adagrad. This was further built upon by Adam (Kingma and Ba, 2014).

### Momentum

If we trust our gradient approximations, there is no reason for us to believe that the gradients move in a direction that is too drastic. We might want to smooth out our route through the space. The gradient that was evaluated from the previous mini-batch might not be completely useless in the current phase and considering a *momentum* in that direction is useful (Polyak, 1964). This avoids unwanted sharp motion in wrong directions due to wrong gradient approximations because of a noisy mini-batch. If the direction of motion is continuous, momentum encourages it, and if the direction of motion changes drastically, momentum will discourage that change. If properly used, this may even accelerate the learning in the generalized direction that carries the most momentum, which generally builds up over a few iterations in the common direction.

Consider building up the change in velocity of motion through the space in one direction by subtracting from a decaying version of an old averaged velocity $v$, a weighted version of the new gradient:

$$v_i^\tau = \alpha^\tau v_i^{\tau-1} - \eta^\tau e'(w_i^\tau) \tag{2.29}$$

Now that we have updated the velocity with the current weight of the gradient, we can indirectly update the weights using this velocity:

$$w_i^{\tau+1} = w_i^\tau + v_i^\tau \tag{2.30}$$

Here, $\alpha$ is kept reasonably close to 1 as we trust the velocity as it builds up over the epochs, but we shall start with a sturdy 0.5 allowing for less impact on the earlier iterations where learning is generally tumultuous. $\alpha$ is basically the trust that we have over the velocity.

Another commonly used momentum technique is Nesterov's accelerated gradient (NAG) method that updates the weights with the velocity first and then accumulates the velocity later on (Nesterov, 1983). The update step for NAG is

$$w_i^{\tau+1} = w_i^\tau + \alpha^\tau v_i^\tau - \eta^\tau e'(w_i^\tau) \tag{2.31}$$

Momentum and second-order updates have drastically changed the speed and stability with which learning takes place in deep neural networks and CNNs. The importance of momentum is discussed in the cited article and will prove to be an interesting read for a curious reader (Sutskever et al., 2013).

Summary

In this chapter, we studied linear regression and the modeling of continuous valued labels. We started with the basics of supervised machine learning and created a simple supervised learning dataset. We made an assumption of linearity with our dataset. With the assumption of linearity, we were able to make our problem simple. We used the linearity assumption to create a linear model and parameterized it. We arrived at an analytic solution for our parameters using the least squares method. We also arrived at an MLE interpretation for our least squares solution.

We applied various constraints on the parameters of our model and studied regularized and ridge linear regression. Regularization makes our learning stable and avoids explosion of weights. The additional constraint terms got added on to our already existing analytical solution and serendipitously, we found that a regularized regression might be easier to solve.

We also studied some potential problems that might arise with these linear models. If the dataset is not linear and our assumption will not hold, we seek a solution that is nonlinear. Basis function expansion helped us in projecting the features onto a nonlinear space and solves a linear regressor on that nonlinear space. This enabled us to solve nonlinear regression using the techniques we learned in applying linear regression. We also studied the problems of overfitting and introduced regularization as one way of alleviating overfitting. We studied $L_1$ and $L_2$ norms and their effect on the smoothness of the learned models. We also studied better evaluation and model averaging using cross-validation.

Not all regression problems have an analytic solution. Large amounts of data and noninvertible matrices in the analytical solution means that the analytical solution, even if it exists, is not practical to use. In this chapter, we also expanded upon our earlier understanding of linear and nonlinear regression by solving the problem without an analytic solution. We used the techniques of optimization, in particular the gradient descent, to solve the problem and find optimal weights. To use our data in smaller batches, we studied the use of stochastic gradient descent.

We studied the effects of varying learning rates during optimization. Using adaptive learning rates that change depending on which stage of the optimization task provided a more stable way to arrive at the local minima. We then moved on to study Adagrad, Newton's method, and other second-order learning techniques. We also studied the geometry of regularization and how different regularizers affected the region in which we let our weights move while optimizing. Another way to control and move weights in a structured fashion were the use of momentum and we studied various forms and types of momentum including the Polyak's and Nesterov's momentum techniques.

# References

Boyd, Stephen and Vandenberghe, Lieven. 2004. *Convex Optimization*. Cambridge: Cambridge University Press.

Button, Katherine S, Ioannidis, John PA, Mokrysz, Claire et al. 2013. Power failure: Why small sample size undermines the reliability of neuroscience. *Nature Reviews Neuroscience* 14: 365–376.

Choromanska, Anna, Henaff, Mikael, Mathieu, Michael et al. 2015. The loss surfaces of multilayer. *AISTATS*.

Duchi, John, Hazan, Elad, Singer, Yoram. 2011, July. Adaptive subgradient methods for online learning and stochastic optimization. *Journal of Machine Learning Research* 12: 2121–2159.

Efron, Bradley and Gong, Gail. 1983. A leisurely look at the bootstrap, the jackknife, and cross-validation. *American Statistician* 37: 36–48.

Felzenszwalb, Pedro F, Girshick, Ross B, and McAllester, David. 2010. Cascade object detection with deformable part models. *2010 Conference on Computer Vision and Pattern Recognition (CVPR)* (pp. 2241–2248). IEEE.

Galton, Francis. 1894. *Natural Inheritance* (Vol. 5). New York, NY: Macmillan and Company.

Golub, Gene H, Heath, Michael, and Wahba, Grace. 1979. Generalized cross-validation as a method for choosing a good ridge parameter. *Technometrics* 21: 215–223.

Gunn, Steve R. 1998. *Support vector machines for classification and regression.* ISIS technical report.

Hinton, Geoffrey, Srivastava, Nitish, and Swersky, Kevin. n.d. *Overview of mini-batch gradient descent.* Retrieved from https://www.coursera.org/learn/neural-networks

Hoerl, Arthur E and Kennard, Robert W. 1970. Ridge regression: Biased estimation for nonorthogonal problems. *Technometrics* 12: 55–67.

Jefferys, William H and Berger, James O. 1992. Ockham's razor and Bayesian analysis. *American Scientist* 80: 64–72.

Kingma, Diederik and Ba, Jimmy. 2014. *Adam: A method for stochastic optimization.* arXiv preprint arXiv:1412.6980.

Kohavi, Ron. 1995. A study of cross-validation and bootstrap for accuracy estimation and model selection. *Proceedings of the 14th International Joint Conference on Artificial Intelligence*, pp. 1137–1145.

Lawson, Charles L and Hanson, Richard J. 1995. *Solving Least Squares Problems* (Vol. 15). Philadelphia, PA: SIAM.

Le Cessie, Saskia and Van Houwelingen, Johannes C. 1992. Ridge estimators in logistic regression. *Applied Statistics* 41(1): 191–201.

Marquardt, Donald W and Snee, Ronald D. 1975. Ridge regression in practice. *American Statistician* 29(1): 3–20.

Nadaraya, Elizbar A. 1964. On estimating regression. *Theory of Probability and Its Applications* 9: 141–142.

Nesterov, Yuri. 1983. A method of solving a convex programming problem with convergence rate O (1/k2). *Soviet Mathematics Doklady* 27: 372–376.

Neter, John, Kutner, Michael H, Nachtsheim, Christopher J et al. 1996. *Applied Linear Statistical Models* (Vol. 4). Pennsylvania Plaza, NY: McGraw-Hill /Irwin.

Polyak, Boris Teodorovich. 1964. Some methods of speeding up the convergence of iteration methods. *USSR Computational Mathematics and Mathematical Physics* 4: 1–17.

Ramsay, James O. 2006. *Functional Data Analysis*. Hoboken, NJ: Wiley Online Library.

Rumelhart, David E, Hinton, Geoffrey E, and Williams, Ronald J. 1986. Learning representations by back-propagating errors. *Nature* 323: 533–538.

Seber, George AF and Lee, Alan J. 2012. *Linear Regression Analysis* (Vol. 936). Hoboken, NJ: John Wiley & Sons.

Shao, Jun. 1993. Linear model selection by cross-validation. *Journal of the American Statistical Association* 88: 486–494.

Shawe-Taylor, John and Cristianini, Nello. 2004. *Kernel Methods for Pattern Analysis*. Cambridge: Cambridge University Press.

Sutskever, Ilya, Martens, James, Dahl, George E et al. 2013. On the importance of initialization and momentum in deep learning. *International Conference on Machine Learning*, Vol. 28, pp. 1139–1147, Atlanta, Georgia.

Vapnik, Vladimir, Golowich, Steven E, Smola, Alex et al. 1997. Support vector method for function approximation, regression estimation, and signal processing. *Advances in Neural Information Processing Systems*, pp. 281–287.

Zhang, Ping. 1993. Model selection via multifold cross validation. *Annals of Statistics* 21: 299–313.

# 3

# ARTIFICIAL NEURAL NETWORKS

The previous chapters have introduced several types of methods for classifying and regressing, given some datasets, often in terms of multidimensional feature vectors. In this chapter, we turn to a different approach: artificial neural networks (ANNs). ANNs are at the root of many state-of-the-art deep learning algorithms. Although a few of the ANNs in the literature were fundamentally motivated by biological systems and some even came with hardware implementations, such as the original McCulloch–Pitts neuron (McCulloch et al., 1943), the vast majority were designed as simple computational procedures with little direct biological relevance. For the sake of practicality, the presentation of this chapter will mostly concern mostly the algorithmic aspects of ANNs.

Consider again the regression problem from Chapter 2, for the dataset $D$ of size $n$, $d$-dimensional $\boldsymbol{x}_i$, and their corresponding one-dimensional label/target value $y_i$, that is, $D = \{(\boldsymbol{x}_i, y_i), i \in [1, 2, \ldots, n]\}$. This linear regression problem can also be schematically illustrated as a *feedforward network* as shown in Figure 3.1. This is a simple linear neuron. In the figure, each circle represents a processing unit, and in the case of linear regression as presented in Chapter 2, all the processing units are linear. Each unit on the *first layer* (all the processing units connected to the data $x_i$) is simply an identity function. The *output* is a single unit connected to $y_i$ (with its input being a simple summation of all those passed in by the incoming edges). The directed edges are annotated by the corresponding weights to be applied (multiplied) to the data passing through the edges. While this is simply another way of presenting the same solution as we have seen before, it is instructive to use this functional structure as it will handily lead to more complex processing paradigms we present later in this chapter.

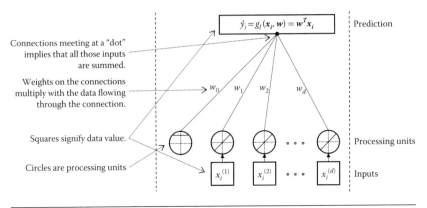

**Figure 3.1**    Illustrating linear regression as a feed-forward linear neuron.

In Figure 3.1, we used the term *feed-forward linear neuron* since the processing units are connected in such a way that their connections go only one way from the input to the output. This will be primarily the type of network we are concerned with in this chapter, although it is worth noting that there are networks that involve feedback paths in the data processing flow and are thus not merely feed-forward.

### The Perceptron

Chapter 2, on regression and optimization, discussed how to model an input–output relation and solve for the optimal model parameters. In a majority of visual computing applications, the task is to perform supervised classification, such as through support vector machines (Gunn, 1998; Vapnik et al., 1997). Interestingly, all these may be viewed as some variant of the simple illustration of Figure 3.1. Let us first consider a binary classification problem, where the label $y_i$ takes only one of two distinctive values, 0 or 1. That is, the dataset $D$ to be learned from has $y_i \in [0,1]$. We use 0 or 1 here as the actual values for the binary $y_i$, although it is possible to use other binary values. Note that while we may still force a linear regression model onto the data without considering the binary nature of the labels, it is unlikely to produce a useful or strong model for future predictions. A natural modification one would like to introduce to the model is to make the output go through a thresholding function. To be precise, we want to model the output as

$$g_P(x_i, w) = \begin{cases} 1, & \text{if} \quad w^T x_i > 0 \\ 0, & \text{otherwise} \end{cases} \tag{3.1}$$

where, for convenience, we have assumed a cutoff to occur at $w^T x_i = 0$. As will become clear later, there are different ways of defining such a cutoff for a processing unit. This thresholding function is some-times called the *activation function* as its output signifies whether the node, with its given inputs, is activated (emitting 1). In our earlier case, we had no activation function or rather the activation function was identity. Note that the activation function performs only thresholding; $w^T x_i$ is created by the summing node. The activation function can be independently defined as

$$a_P(t) = \begin{cases} 1, & \text{if} \quad t > 0 \\ 0, & \text{otherwise} \end{cases} \tag{3.2}$$

In this case, $t$ is the linear regression output on top of which we build an activation (thresholding) function.

The resultant neuron essentially illustrates the perceptron or a *linear threshold neuron*. This model is similar to the one first proposed by Rosenblatt, as an alternative and improvement to the simpler McCulloch–Pitts neuron model and as a means to use the linear neuron to perform classification (Rosenblatt, 1958) as shown in Figure 3.2. This and several other variants of the perceptron along with their geometry in the feature spaces were studied thoroughly in 1960s, leading to the first wave of ANNs and their becoming relatively dormant soon after their limitations were identified (Minski et al., 1969).

The original algorithm that was used to train this network was often referred to as the perceptron algorithm. The perceptron algorithm is a useful algorithm that ensures that one can learn the weights of the perceptron so as to get a good decision boundary. It is also an online learning algorithm, meaning it can learn sample by sample. We present the Rosenblatt version of the online perceptron algorithm in Figure 3.3.

There have been several proofs and studies regarding the convergence/performance of this seemingly simple-looking algorithm, especially when the input samples are linearly separable, from as early as 1960s (Novikoff, 1962) to recent years (Mohri and

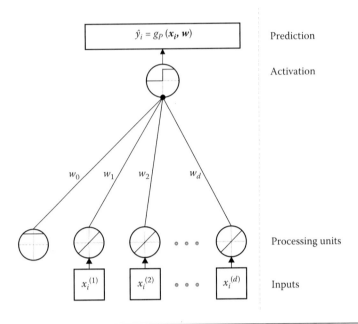

**Figure 3.2**  Illustration of a Rosenblatt perceptron.

**Input**

 Training set $D = \{ (x_i, y_i), i \in [1,2,...n]\}, y_i = [0,1]$.

**Initialization**

*Step* 1: Initialize the weights and threshold for the network.
*Step* 2: Weights may be set to 0 or small random values.

**Iterate**

   **for** $t \to$ convergence
   {
      **for** each sample $x_i$ with label $y_i$:
      {
         *Step* 1: Compute the output $\hat{y}_i$ of the network.
         *Step* 2: Estimate the error of the network $y_i - \hat{y}_i$.
         *Step* 3: Update the weight $w(t + 1) = w(t) + e(w)x_i$.
      }
   }

**Figure 3.3**  The perceptron algorithm.

Rostamizadeh, 2013). An interested reader may refer to such literature for more elaborate discussion.

 It is interesting to note that, at about the same time when the original perceptron was introduced, a regression model, the logistic regression, was developed by David Cox in 1958 (Cox, 1958). In the basic logistic regression, the key task is to estimate the probabilities of the dependent variable $y$ taking one of the binary values 0 or 1, using the logistic function. That is, we model $P(y = 1|x) = g_\sigma\{x, w\}$ with the function $g_\sigma$ defined as

$$P(y = 1|x) = g_\sigma\{x, w\} = \frac{1}{1 + e^{-w^T x}} \tag{3.3}$$

The activation function here is

$$a_\sigma(t) = \frac{1}{1 + e^{-t}} \tag{3.4}$$

The task is to find the maximum-likelihood solution for $w$, given the training data.

In general, the maximum-likelihood estimation cannot be solved in a closed form and some optimization methods like those discussed in Chapter 2 should be employed to develop an iterative procedure by updating some initial guess for $w$, with the objective function of the optimization naturally being the log-likelihood (or negative log-likelihood, for minimization). We will omit the discussion of such procedures as they can be viewed as generalizations from the discussion on gradient descent from Chapter 2. An interested reader may also check the relevant literature, for example, Hosmer Jr et al. (2004), for more information since they are not the focus of this discussion.

Note the similarity between the functions $g_\sigma(.)$ and $g_P(.)$; they are both some thresholding functions. $g_\sigma(.)$ may be viewed as a soft-thresholding function, as evident from a plot of the function in Figure 3.4. We may use it to replace the hard threshold unit in the

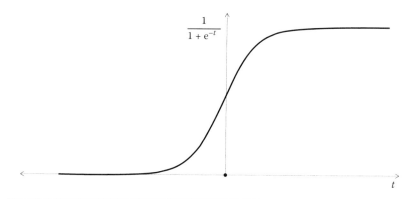

**Figure 3.4**   Illustration of the standard logistic function.

perceptron of Figure 3.2, resulting in the illustration of Figure 3.5. The benefit in doing so is that the logistic function is continuous and differentiable in its entire support. Differentiability is an important property that lends us now the means to establish an error gradient such as what we did for linear regression in Chapter 2. We may use all the tools and techniques of the gradient descent method and other optimization methods that we studied in Chapter 2 to iteratively search for a solution for $w$.

Similar to our linear regression problem, we can define a loss function $e(w)$ and update the weights using gradient descent as

$$w^{\tau+1} = w^{\tau} - \eta^{\tau} e'(w^{\tau}) \tag{3.5}$$

where $\eta^{\tau}$ is a scalar controlling how much adjustment we make to the weight vector during each iteration and, as was the case with the previous problem, is the *learning rate*.

While the above introduction of the logistic function into the neuron has not fundamentally changed the basic idea of the perceptron algorithm as we only slightly modified the procedure for updating the weights, moving from the perceptron to gradient

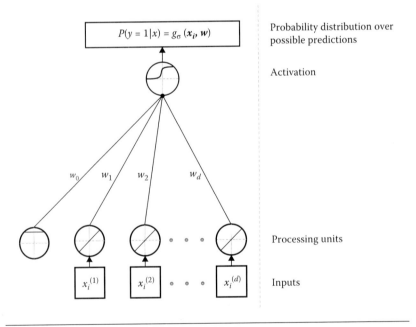

**Figure 3.5**  Illustration of a logistic neuron.

descent is a major improvement computationally. The benefit of using the logistic function or other similar continuous and differentiable functions for activation is that they enable a more principled way of updating the weights when we extend the learning to multilayer networks. This leads to the backpropagation (BP) algorithm (Bryson, 1961; Kelley, 1960), which we shall build upon and study through the rest of this book. The importance of the BP algorithm cannot be understated in any study involving deep learning and neural networks.

Also, in general, since we may now define some loss functions that are not necessarily directly based on binary classification, the goal of learning should be ideologically considered as finding a solution that minimizes the defined loss. This is more flexible in many real applications, as will be demonstrated by more recent deep learning-based applications with various loss functions that will be introduced in subsequent chapters.

Let us further extend the basic single neuron model of Figure 3.5 into one with more than one output node. This extension is illustrated in Figure 3.6. Besides helping to set up the notation properly for later discussion, the extension is necessary if we wish to handle more than two classes. In general, if we have $c$ classes to process, we would employ $c$ output nodes, as illustrated in the figure.

The output of this network is not directly a prediction of the labels but a probability distribution over the classes. The prediction $\hat{y}$ is simply the class that gives the maximum of such probabilities. Considering there are $c$ classes, we have $c$ outputs that give us a vector of outputs at the top of the dot-product step before the softmax layer:

$$l_1 = W^T \cdot x_i \qquad (3.6)$$

where $W$ is a matrix of size $d \times c$ such as

$$W = \begin{bmatrix} w_0^{(1)} & w_0^{(2)} & \cdots & w_0^{(c)} \\ w_1^{(1)} & w_1^{(2)} & \cdots & w_1^{(c)} \\ \vdots & \vdots & \ddots & \vdots \\ w_d^{(1)} & w_d^{(2)} & \cdots & w_d^{(c)} \end{bmatrix} \qquad (3.7)$$

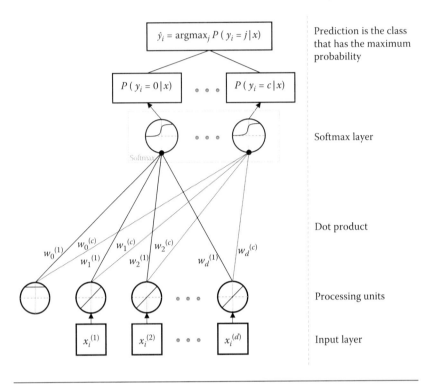

**Figure 3.6**  A network with more than one output node and the notations.

Taking the dot product such as the one shown in Equation 3.6 will yield us a vector of $l_1$ such as $l_1 = [l_1^{(1)}, l_1^{(2)}, ..., l_1^{(c)}]$, where $c$ is the number of classes. The subscript 1 represents the output for the first layer. Here we only have one layer. These are not normalized vectors, therefore they are not probabilities. To produce a probability distribution over all classes, we need to normalize these. We can make use of the *softmax* function to do this normalization. The softmax-activated output of the softmax layer is

$$P(y_i = j \mid x_i) = \sigma(l_1^{(j)}) = \frac{e^{l_1^{(j)}}}{\sum_{k=1}^{c} e^{l_1^{(k)}}}, \quad j \in [1, 2, ..., c] \qquad (3.8)$$

The predicted class is the class that maximizes this probability:

$$\hat{y}_i = \text{argmax}_j \, P(y_i = j \mid x_i) \qquad (3.9)$$

Thus, we have a network that predicts the class for a multiclass classification problem as illustrated in Figure 3.6. Let us create a log-likelihood for this network like what we created for linear regression. Considering we have the probabilities $P(y_i = j \mid x)$ for all classes $j \in [1, 2, \ldots, c]$:

$$l(W) = \frac{1}{n} \sum_{i=1}^{n} \sum_{j=1}^{c} \mathbb{I}_j(y_i) \log(P(y_i = j \mid x_i)) \qquad (3.10)$$

where

$$\mathbb{I}_j(y_i) = \begin{cases} 1, & \text{if } y_i = j \\ 0, & \text{otherwise} \end{cases} \qquad (3.11)$$

is an indicator function that indicates the class. The indicator function is 1 only for one of the classes and is 0 for the others. This means that in maximizing the log-likelihood, what we are doing in essence is maximizing the probability of the correct class. We can now perform gradient descent on the negative of $l(W)$, which is the negative log-likelihood of the logistic regressor. The gradients involved and the procedure to perform gradient descent on this gradient are a slight extension from Chapter 2 and are thus left to the reader. We will go into more detail later in the chapter when we study the backpropagation algorithm. Note that Equation 3.10 is the average likelihood of the entire dataset. This averaging enables us to use mini-batches that can have similar learning procedures regardless of the amount of data used. This also enables us in using similar learning rates regardless of the data batch size. Even if we were to perform gradient descent using stochastic gradient descent, our learning rates do not have to depend on batch size as the likelihood itself is normalized by the number of samples in the batch $n$.

So far we have studied various types of neurons including the perceptron, linear regressor neuron, and the softmax perceptron. Various types of problems could be forced onto these neurons. For instance, a max-margin neuron to produce a max-margin or a support vector machine (SVM)-type classifier could be produced by using the following likelihood:

$$l(W) = \max(0, 1 - y \cdot \hat{y}) \qquad (3.12)$$

To use this likelihood, we need to convert the labels from $[0, 1]$ to $[-1, 1]$. If our prediction matches the true value (or is close to the true value), we get the likelihood as 0; if our prediction was wrong, we get a likelihood that is large (with a worst case of 2). Note that many types of neurons exist and we have deliberately limited ourselves in this book to studying those that are popular and relevant for computer vision. There is a plethora of research papers available in the literature that could quench the reader's thirst for material on this.

## Multilayer Neural Networks

The perceptron algorithm and logistic regression produce linear decision boundaries and thus if the two classes are linearly separable, the algorithms may be used to find a solution. However, a single-layer perceptron cannot deal with cases that are not linearly separable. A simple example of this is the XOR problem, which is illustrated in Figure 3.7. There is no linear solution to solve this problem. We may employ basis function expansions as we saw in the previous chapter and may be able to solve the problem using a higher order transformation of the data (Powell, 1977). This further raises the question of what basis functional space to choose to project the features onto so that we can obtain good linear separation in that space.

One way to solve this problem is by using a multilayer neural network (MLNN). Using MLNNs, we will be able to *learn* such a functional transformation. Often in computer vision, scientists talk of *feature extraction*. We introduced feature extraction briefly in

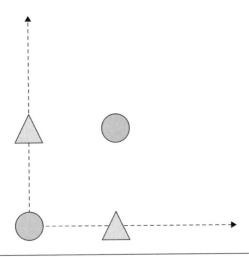

**Figure 3.7**   Illustrating the XOR problem in the two-dimensional feature space. Triangles and circles represent two different classes in the feature space.

Chapter 1 and discussed a few handcrafted features that are popular in computer vision. In a sense, these features are nonlinear transformations of the original pixels. Typically, such features are designed and tuned by domain experts. The hope is that, after mapping the raw images into the feature domain, analysis tasks such as linear classification can be done more effectively. In practice, since these transformations are fixed procedures that are designed by hand, they are in general not adaptable to different problems that may utilize different loss functions.

Using MLNNs, we will be able to learn such transformations. We will see that with deeper networks, we can learn much more sophisticated feature spaces where indeed we will be able to perform linear classification.

In the basic perceptron, the processing unit is linear. This appears to suggest that one may be able to gain the capability of producing nonlinear decision boundaries if the units can perform nonlinear processing. There are different possibilities for bringing nonlinearity into the processing units. For example, the radial basis function network achieves this by introducing some nonlinear radial basis functions (Lowe and Broomhead, 1988). Another approach would be to simply introduce at least one layer of nonlinear units with adjustable weights into the basic perceptron. This results in an MLNN, and owing to its traditional fraternity to the perceptron, such a network is also often referred as a *multilayer perceptron* (MLP). In the context of this book and in most literature, with some abuse of terminology, we often use these two terms interchangeably. The middle layer of the simple MLNN shown in Figure 3.8 is *hidden* from both the input $x$ and the output $y$, and thus it is referred to as a hidden layer. The hidden layer has $d_2$ nodes as opposed to $d$ from the input layer that gets its dimensionality from the number of dimensions of the input itself.

It is pleasantly surprising how this simple trick of stacking the single-layer perceptron could lead us to a solution to the XOR problem shown in Figure 3.7 and how it could learn an effective feature mapping. Even without going into detail, one can notice in Figure 3.8 that the classifier $g$ works on a transformed version of the inputs that is defined by $l_1$. It is also clear that this transformation to $l_1$ is dependent on the learnable parameters $w^2$. This thus gives us an intuitive picture of how a network

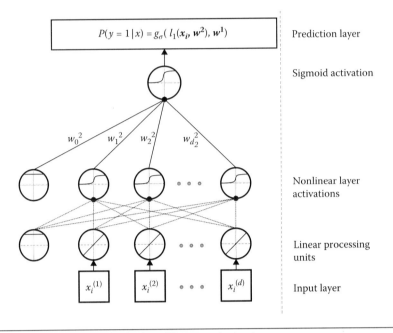

**Figure 3.8**    Illustration of a simple multilayer neural network.

with a single hidden layer could learn its own internal mapping to a new feature space. Each node on the hidden layer can also be thought of as a logistic regression neuron since it adds up the signals and performs a nonlinear activation on it. Thus, we have stacked an entire layer of neurons on the hidden layer before it is fed to the regressor.

Before we discuss how to learn the parameters of this new multi-layer network, we demonstrate the existence of a solution to the XOR problem in Figure 3.9, for a network with two nodes and a bias in the hidden layer. In the hidden layer, the node on the left with ∧ by its side will not activate unless both the inputs are 1. This acts as an AND gate by itself. The node on the right with ∨ by its side activates even if one of the signals is 1. This acts as an OR gate by itself. The second layer builds up the XOR. The feature space and classification regions are also shown in Figure 3.9. Thus, we are able to demonstrate that the XOR solution indeed exists with a neural network with one hidden layer. A solution close to this could be learned using a simple algorithm called backpropagation that we discuss in the next section. Given this solution for this architecture, one node in the hidden layer will always act as an AND and the other will always act as an OR in order to

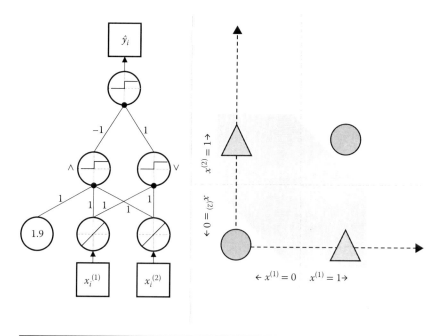

**Figure 3.9** An MLNN that solves the XOR problem in the two-dimensional space.

produce XOR properties. The reader is encouraged to try the same problem with more than two nodes in the hidden layer, find other solutions, and observe the effects. Similar to what we have seen earlier, in general, we may have more than one output node when dealing with a multiclass problem. Further, we may have multiple hidden layers too, and it is worth noting at this moment that many recent deep learning architectures employ more than a few hidden layers.

In the literature, there are different ways of counting the number of layers in an MLP. Consider the XOR network in Figure 3.9. Sometimes, the network is said to have three layers (corresponding to the layers of processing units), while sometimes it said to have two layers as the input layer is trivial. We will use the first way of counting the layers and will use the notation of an "$n_I \rightarrow n_H \rightarrow n_C$ MLP" to denote a three-layer network with $n_I$ input nodes, $n_H$ hidden nodes in the first hidden layer, and $n_C$ output nodes.

Using the above three-layer MLP as an example, we are now ready to define mathematically the general input–output relation for an arbitrary $n_I \rightarrow n_H \rightarrow n_C$ feed-forward network:

$$\hat{y} = a_C(a_H(a_I(\boldsymbol{x}), \boldsymbol{w}_H), \boldsymbol{w}_C) \tag{3.13}$$

Note that we have intentionally used different activation functions, $a_C$, $a_H$, and $a_I$ for the output, hidden, and input layers, respectively, just to illustrate the possibility that we can allow different activation functions across layers and that even the input layer can have an activation. In fact, it is even possible to allow each of the hidden or output nodes to assume their own distinctive activation functions too, although in practice it is more customary to use the same function for all the nodes in the same layer (and sometimes for all the nonlinear layers) for convenience of implementation.

Despite the simplicity in its form, the above equation of the input–output relation for a three-layer network turned out to be very general: so general that *in theory* it may be used to approximate any continuous function from the input to the output, provided we have a sufficient number of hidden units $n_H$, proper nonlinearities, and that proper weights could be defined. This is deemed by some as proven largely due to a theorem by Kolmogorov (1956), although the relevance of the Kolmogorov theorem to the multilayer network has also been disputed largely for reasons of *practicality* (Girosi and Poggio, 1989).

Rigorous theoretical proof aside, we may still be able to appreciate why a three-layer network would work so well for mapping any input–output relation, using some intuitive interpretations. For example, from a Fourier analysis perspective as discussed in Chapter 1, any continuous function can be approximated to arbitrary accuracy by a possibly infinite number of harmonic functions (or the Fourier bases), and one may imagine the hidden nodes may be tuned to simulate the harmonic functions so that the output layer can draw upon the hidden nodes to form a desired approximation to a function. By invoking the simple relation between regression and the perceptron illustrated in Figures 3.1 through 3.4, we may relate the MLP to *projection pursuit regression* since the hidden layer can act as a projection of the input to a *typically* much lower dimensional space where the classification (or regression) is done (Friedman and Stuetzle, 1981; Lingjaerde and Liestol, 1998). It is interesting to note that, in both these arguments, it is assumed that the learning of the network is to figure out some good "representations" for the input so as to facilitate the later decision-making, much like the ideas we have been trying to accomplish through neural networks.

We now turn to the problem of training the network parameters with a given dataset. It is not difficult to realize that we could simply handle the output layer weights by following the same principle in training the single-layer perceptron. Based on any predefined loss function, we first compute the loss for the current input or input batch under the current network's feed-forward mode. Then the weights of the output nodes can be adjusted using any of the optimization methods we studied in the previous chapter. Let us consider the gradient decent method. We notice that in gradient descent the loss was directly a function of these weights. The same cannot be said for the hidden layer nodes in these networks, since a given training set would not contain any "target labels" for a hidden node. The training data do give target labels for the output nodes but that does not give us a target for the hidden nodes. Since the hidden nodes are connected to the output layer that is supervised, with proper target labels and thus computable loss, the key task then would be to properly relay the loss computed in the output layer back to each of the hidden nodes.

### The Back-Propagation Algorithm

The back-propagation (BP) algorithm has been the primary method for achieving the learning of the parameters of an MLP. The development of the BP algorithm in the literature was quite gradual and the invention of this method is not attributed to any single article or group of authors. It appears though that one article in the late 1980s brought to attention the importance and significance of this algorithm (Rumelhart 1988). The core idea of the BP algorithm for gradient-decent-based weight learning is quite simple. It basically relies on the chain rule of differentiation for making a connection between the loss computed at the output layer and any hidden nodes. This connection helps to relay the final loss of the network back to any earlier layers/nodes in the network so that the weights of those layers/nodes may be proportionally adjusted (to the direction that reduces the loss). Formally, we now present the BP algorithm, using the network of Figure 3.10 for illustration.

To use this idea of gradient descent for any network parameter $w$, we need to first find the gradients of the error $e$, with respect to that parameter. Essentially, we need to be able to calculate $\dfrac{\partial e}{\partial w_{i,j}^{(k)}}$, $\forall$ $w_{i,j}^{(k)}$ in the network architecture for the $i^{\text{th}}$ layer, $j^{\text{th}}$ neuron and its $k^{\text{th}}$

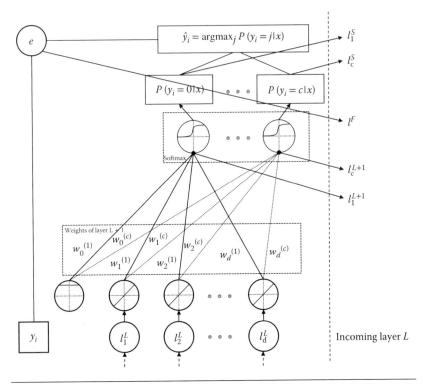

**Figure 3.10** A feed-forward network used for illustrating the back-propagation algorithm. In this cut-away diagram we show the last layer in action along with the softmax layer and the error layer for the $i^{th}$ sample.

dimension. Ignoring the $i$-layer sub-script as we only consider one parametrized layer, we will show how to derive this derivative across one layer using the chain rule. This may be extended beyond one layer by simple extension. The term $l^F$ standing for the final layer's outputs and is defined in Equation (3.10). Terms $l_j^S$ are defined in Equation (3.8), which are the softmax outputs. The terms $l_j^{L+1}$ are the outputs of a dot-product defined by $l_j^{L+1} = \sum_i w_i^{(j)} . l_i^{(L)}$, ignoring the layer id on the weight notations. Terms $l_j^L$ are simply the incoming layer inputs coming from a previous layer. What we seek here are the terms $\dfrac{\partial e}{\partial w_j}$.

Once we derive this for some $j$, we can follow the same procedure for all $j$. Once we can do this for this cut-away picture, we can simply extend it for all layers.

Consider again the error function using this notational setup,

$$
e(\boldsymbol{w}) = l^F \left[ \begin{array}{l} l_1^S \left\{ l_1^{L+1} \left( \left\langle l_1^{(L)}, w_1^{(1)} \right\rangle, \ldots \left\langle l_d^{(L)}, w_d^{(1)} \right\rangle \right) \right\}, \ldots \left\{ l_c^{L+1} \left( \left\langle l_1^{(L)}, w_1^{(c)} \right\rangle, \ldots \left\langle l_d^{(L)}, w_d^{(c)} \right\rangle \right) \right\}, \\ \ldots l_c^S \left\{ l_1^{L+1} \left( \left\langle l_1^{(L)}, w_1^{(1)} \right\rangle, \ldots \left\langle l_d^{(L)}, w_d^{(1)} \right\rangle \right) \right\}, \ldots \left\{ l_c^{L+1} \left( \left\langle l_1^{(L)}, w_1^{(c)} \right\rangle, \ldots \left\langle l_d^{(L)}, w_d^{(c)} \right\rangle \right) \right\} \end{array} \right] \tag{3.14}
$$

The nested nature of this error depicts the modular nature of the layers used. In this setup, for any weight vector $\boldsymbol{w}_k$, the partial derivative $\dfrac{\partial e}{\partial \boldsymbol{w}_k}$ can be derived as,

$$
\frac{\partial e}{\partial \boldsymbol{w}_k} = \sum_{i=1}^{c} \sum_{j=1}^{c} \frac{\partial l^F}{\partial l_i^S} \frac{\partial l_i^S}{\partial l_j^{L+1}} \frac{\partial l_j^{L+1}}{\partial \boldsymbol{w}_k} \tag{3.15}
$$

Consider this summation expanded for the simple case of $c = 2$, we get the following,

$$
\frac{\partial e}{\partial \boldsymbol{w}_k} = \frac{\partial l^F}{\partial l_1^S} \frac{\partial l_1^S}{\partial l_1^{L+1}} \frac{\partial l_1^{L+1}}{\partial \boldsymbol{w}_k} + \frac{\partial l^F}{\partial l_1^S} \frac{\partial l_1^S}{\partial l_2^{L+1}} \frac{\partial l_2^{L+1}}{\partial \boldsymbol{w}_k} +
$$
$$
\frac{\partial l^F}{\partial l_2^S} \frac{\partial l_2^S}{\partial l_1^{L+1}} \frac{\partial l_1^{L+1}}{\partial \boldsymbol{w}_k} + \frac{\partial l^F}{\partial l_2^S} \frac{\partial l_2^S}{\partial l_2^{L+1}} \frac{\partial l_2^{L+1}}{\partial \boldsymbol{w}_k} \tag{3.16}
$$

We can already notice terms that are repeating in this summation and would not require to be calculated several times over. This provides us an optimal way to calculate gradients efficiently. Let us re-write Equation (3.16) as,

$$
\frac{\partial e}{\partial \boldsymbol{w}_k} = \frac{\partial l_1^{L+1}}{\partial \boldsymbol{w}_k} \left[ \frac{\partial l^F}{\partial l_1^S} \frac{\partial l_1^S}{\partial l_1^{L+1}} + \frac{\partial l^F}{\partial l_2^S} \frac{\partial l_2^S}{\partial l_1^{L+1}} \right] +
$$
$$
\frac{\partial l_2^{L+1}}{\partial \boldsymbol{w}_k} \left[ \frac{\partial l^F}{\partial l_1^S} \frac{\partial l_1^S}{\partial l_2^{L+1}} + \frac{\partial l^F}{\partial l_2^S} \frac{\partial l_2^S}{\partial l_2^{L+1}} \right], \tag{3.17}
$$

or further as,

$$
\frac{\partial e}{\partial \boldsymbol{w}_k} = \frac{\partial l_1^{L+1}}{\partial \boldsymbol{w}_k} \left[ \Delta(S_1) \right] + \frac{\partial l_2^{L+1}}{\partial \boldsymbol{w}_k} \left[ \Delta(S_2) \right]. \tag{3.18}
$$

The vector $\Delta(\boldsymbol{S})$ containing the terms, $\Delta(S_1)$ and $\Delta(S_2)$ are not dependent on the layer $L + 1$ and can be marginalized out. These are calculated at the layer $S$ using the outputs of layer $L + 1$ only and therefore could be abstracted away and provided as inputs from the top to the layer $L + 1$.

To implement, consider a layer $q$ with parameters $w_q$ as a block of operations, which accepts incoming input vector $l^{q-1}$ and outputs vector $l^q$. It also internally computes a gradient which could store a version of it as, $\dfrac{\partial e}{\partial w_q}$. To do this, all it would need is the gradient vector $\Delta(q)$ from the previous layer. After computing its gradient, it could propagate the gradient $\Delta(q-1)$ to the layer below.

This is the BP algorithm. The inputs are fed forward from bottom to top, where the error is calculated. Layer by layer, from top to bottom, gradients are then propagated until all layers have the error with respect to the gradient calculated. Once we know the gradients of the error with respect to all the weights in the network, we can use the weight update equation of a chosen optimization technique, such as those from the previous chapter, to modify the weights iteratively, hence achieving learning.

### Improving BP-Based Learning

Over many years of development, various "tricks" have been proposed to improve the above basic learning protocol for the multilayer network with some more theoretically motivated and some others inspired by pragmatisms. We have already studied some of these such as momentum, modified learning rates, and second-order methods in the context of optimization in Chapter 2. In the following, we highlight some more of those developments that have been widely used, and many of them that have influenced various recent techniques reported in the current deep learning literature, as will be elaborated in the subsequent chapters.

#### Activation Functions

While we have only discussed the thresholding, identity, and softmax activation functions, there are several other activation functions that are more relevant in the recent literature; we shall study some of them here.

Some other well-studied activations include the squared function and the tanh functions. These have not proved helpful due to practical learning and stability-related issues. One of the most commonly used modern activation functions is the rectifier or *the rectified linear unit*

(ReLU). The rectifier was first introduced in the early 2000s and was later reintroduced in 2010 to great success (Hahnloser et al., 2000; Nair and Hinton, 2010). The rectifier is of the following form:

$$a_{\text{ReLU}}(t) = \max(0, t) \qquad (3.19)$$

This is not a smooth function. The analytical form of an actual implementation of the rectifier is

$$a'_{\text{ReLU}}(t) = \ln(1 + e^x) \qquad (3.20)$$

which is a smooth approximation to the rectifier and is often called softplus.

The plots of these activations are shown in Figure 3.11. As can be observed in Figure 3.11, ReLU has a much steeper profile lending to faster and better learnability (He et al., 2015; Nair and Hinton, 2010).

An extension to the rectifier is the noisy ReLU defined by

$$a_{\text{ReLU}}(t) = \max(0, t + \varphi), \quad \text{with } \varphi \sim \mathcal{N}(0, \sigma(t)) \qquad (3.21)$$

Noisy ReLUs create these random errors in the activation that allow for the activations to be minutely wrong. This allows the network to wander off a little bit while learning. Consider the case where the error surface is full of peaks and valleys. Allowing the parameters to wander around in a restricted fashion helps in probing parts of the space that might not have been accessible using strict gradient

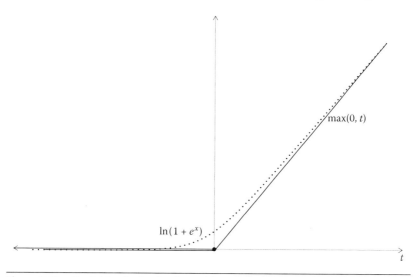

**Figure 3.11** The ReLU activation function and its analytical approximation.

descent. In practice, this may also help us alleviate the problems of overfitting. Noisy ReLUs work particularly well in some computer vision tasks.

A further addendum to ReLUs is the leaky ReLU (Maas et al., 2013). Leaky ReLUs are of the form:

$$a_{\text{ReLU}}(t) = \begin{cases} t, & \text{if } t > 0 \\ \delta t, & \text{otherwise} \end{cases} \tag{3.22}$$

where the parameter $\delta$ is a small constant. This allows the neuron to be active very mildly and produces a small gradient no matter whether the neuron was intended to be active or not. A further development on this is the parametric leaky ReLU, where $\delta$ is considered another parameter of the neuron and is learned along with the BP of the weights themselves. If $\delta < 0$, we obtain an interesting activation function shown in Figure 3.12. In this case, the neuron quite literally produces a negative signal. This makes the gradients move much faster even when the neuron is not contributing to the class predictions.

Another recent activation function is the maxout (Goodfellow et al., 2013). The maxout is an activation function that is quite general so that it can learn its form. It can simulate any activation function from a linear rectifier to a quadratic. Maxout is of the following form:

$$a_{\text{maxout}}(t_i) = \max_{j \in [1,k]} t_{i,j} \tag{3.23}$$

Maxout considers the neighboring $k$ nodes' outputs and produces the maximum of those outputs as the activation. Maxout reduces the number of features that are produced by dropping those features that are not maximum enough. In a manner of speaking, maxout assumes (forces) that nearby nodes represent similar concepts and picks only

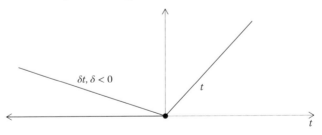

**Figure 3.12**   Leaky ReLU.

one of them, which is the most active. Maxout appears to be an interesting activation function whose performance and value is neither fully understood nor fully experimented upon yet. More research would yield increased understanding of maxout and the reader is encouraged to follow up on this topic.

*Weight Pruning*

Depending on the number of layers of a network and how many nodes each hidden layer will have, a typical MLNN may have a lot of weights, or equivalently a large degree of freedom, resulting in a potentially overcomplicated system. A system that is too complicated suffers from two obvious disadvantages: high computational complexity (and thus difficult to train) and tendency to overfitting (and thus poor generalization performance after training). On the other hand, given a difficult learning task, it is also challenging to precisely determine in advance what would be the optimal size for a low-complexity network that still does an adequate job for the given task. A common approach to alleviating this dilemma is to start with an obviously larger-than-necessary network and then prune the network by deleting the weights and/or nodes to obtain a simpler network.

Generally speaking, the above weight-pruning task may be achieved by two types of approach or their variants. The first type of approach employs some heuristics in selectively removing the weights if they are deemed as having little impact on the final error/cost function. The second type is somewhat subtler in that it relies on introducing additional regularization such as $L_1$ terms in the error function so that smaller weights will be favored (essentially pushing some weights to zero). Over the years, many specific techniques have been developed, employing either or both of the above approaches (Castellano et al., 1997; Lecun et al. 1990; Reed, 1993; Suzuki et al., 2001). It is worth noting that a more recent work also considered how to deal with the potential irregularity of a network that has gone through a weight-pruning process (Anwar et al., 2015).

*Batch Normalization*

Throughout the course of this book, we will study some additional tricks in learning deep networks. Some of these are regularizers such as dropouts, while others are techniques of initialization such as pre-trained networks and mentoring. One of the most powerful and most common among these is batch normalization.

Typically, it is quite common to normalize the images before we feed them forward through a network. Normalization typically involves ranging the image values to [−0.5, 0.5], typically with a mean of 0. In a deep network, the input distribution of each layer keeps varying per batch and per sample. This is because the parameters of the previous layers change during every update. This makes training very difficult particularly with activation functions that saturate.

Although we have assumed thus far in our discussions that all samples from the same class are sampled independently and identically, this is not always true. Samples differ in their statistical properties across batches of data even among the same class. This phenomenon is called covariate shift. To fix this problem of covariate shift, we should normalize the activations coming off every layer. The right variance to normalize with and the mean to mean-subtract the data are often unknown and can be estimated from the dataset itself. Batch normalization is one such way to do it. If $z$ were the activations of one layer, we compute

$$z_{bn} = \frac{(z - \mu_z) * \alpha_z}{\sigma_z} \tag{3.24}$$

where $\mu_z$ and $\sigma_z$ are the mean and the variance of that activation batch, respectively. $\alpha$ is now one of the learnable parameters of the network and can be thought of as learning the stretch of the normalization applied. $\alpha$ is also learned during the same optimization along with the weights. $\alpha$ can be learned for multiple layers using BP. Batch normalization is a powerful tool and helps the network to learn much faster even with nonsaturating activation functions. Batch normalization is particularly popular in visual computing contexts with image data.

### Summary

The discussion in this chapter was intended to introduce feed-forward MLNNs by reviewing key historical developments as well as illustrating the basic models and their notational conventions. We also reviewed some "tricks" that were commonly used during the "second wave" of ANNs that started roughly in the mid-1980s. As such, the presentation was largely focused on the literature during and before that period, although we did cover a few more recent activation functions. It is interesting to note that some of the most recent deep learning techniques, while seemingly emerging only fairly recently in the "third wave" of neural networks, may have deep roots in those earlier developments.

Even if our presentation of the development of MLP in this chapter is brief, it should be evident from the discussion that two key problems are of primary concern in a neural network approach: designing the network architecture (e.g., number of layers, connectivity between layers, choice of activation functions, and number of hidden nodes, etc.) and designing a learning algorithm (possibly employing many of the "tricks" we have illustrated). As will be shown in subsequent chapters, many new efforts in the recent literature report mostly new solutions/designs for these two problems in some specific problem domains.

## References

Anwar, Sajid, Hwang, Kyuyeon, and Sung, Wonyong. 2015. *Structured pruning of deep convolutional neural networks.* arXiv preprint arXiv:1512.08571.

Bryson, Arthur E. 1961. A gradient method for optimizing multi-stage allocation processes. *Proceedings of the Harvard University Symposium on Digital Computers and Their Applications.*

Castellano, Giovanna, Fanelli, Anna Maria, and Pelillo, Marcello. 1997. An iterative pruning algorithm for feedforward neural networks. *IEEE Transactions on Neural Networks* 8(3): 519–531.

Cox, David R. 1958. The regression analysis of binary sequences. *Journal of the Royal Statistical Society. Series B (Methodological)* 20(2): 215–242.

Friedman, Jerome H and Stuetzle, Werner. 1981. Projection pursuit regression. *Journal of the American statistical Association* 76: 817–823.

Girosi, Federico and Poggio, Tomaso. 1989. Representation properties of networks: Kolmogorov's theorem is irrelevant. *Neural Computation* 1: 465–469.

Goodfellow, Ian J, Warde-Farley, David, Mirza, Mehdi et al. 2013. Maxout networks. *International Conference on Machine Learning*, pp. 1319–1327.

Gunn, Steve R. 1998. *Support vector machines for classification and regression.* ISIS technical report.

Hahnloser, Richard HR, Sarpeshkar, Rahul, Mahowald, Misha A et al. 2000. Digital selection and analogue amplification coexist in a cortex-inspired silicon circuit. *Nature* 405: 947–951.

He, Kaiming, Zhang, Xiangyu, Ren, Shaoqing et al. 2015. Delving deep into rectifiers: Surpassing human-level performance on imagenet classification. *Proceedings of the IEEE International Conference on Computer Vision*, pp. 1026–1034. IEEE.

Hosmer Jr, David W and Lemeshow, Stanley. 2004. *Applied Logistic Regression.* Hoboken, NJ: John Wiley & Sons.

Kelley, Henry J. 1960. Gradient theory of optimal flight paths. *ARS Journal* 30: 947–954.

Kolmogorov, Andrey N. 1956. On the representation of continuous functions of several variables as superpositions of functions of smaller number of variables. *Soviet Mathematics Doklady* 108: 179–182.

LeCun, Yann, Denker, John S and Solla, Sara A. 1990. Optimal brain damage. *Advances in Neural Information Processing Systems* 598–605.

Lingjaerde, Ole C and Liestol, Knut. 1998. Generalized projection pursuit regression. *SIAM Journal on Scientific Computing* (SIAM) 20: 844–857.

Lowe, David and Broomhead, D. 1988. Multivariable functional interpolation and adaptive networks. *Complex Systems* 321–355.

Maas, Andrew L, Hannun, Awni Y, and Ng, Andrew Y. 2013. Rectifier non-linearities improve neural network acoustic models. *Proceedings of the 30th International Conference on Machine Learning,* Atlanta, Georgia.

Minski, Marvin L and Papert, Seymour A. 1969. *Perceptrons: An Introduction to Computational Geometry.* Cambridge, MA: MIT Press.

Mohri, Mehryar and Rostamizadeh, Afshin. 2013. *Perceptron mistake bounds.* arXiv preprint arXiv:1305.0208.

Nair, Vinod, and Hinton, Geoffery. 2010. Rectified linear units improve restricted boltzmann machines. *Proceedings of the 27th International Conference on Machine Learning.* Haifa, Israel. 807–814.

Novikoff, Albert BJ. 1962. On convergence proofs on perceptrons. *Symposium on the Mathematical Theory of Automata*, pp. 615–622.

Powell, Michael James David. 1977. Restart procedures for the conjugate gradient method. *Mathematical Programming* 12: 241–254.

Reed, Russell. 1993. Pruning algorithms-a survey. *IEEE Transactions on Neural Networks* 4(5): 740–747.

Rosenblatt, Frank. 1958. The perceptron: A probabilistic model for information storage and organization in the brain. *Psychological Review* 65: 386.

Rumelhart, David E, Hinton, Geoffrey E, and Williams, Ronald J. 1988. Learning representations by back-propagating errors. *Cognitive Modeling* 5: 386–394.

Suzuki, Kenji, Horiba, Isao, and Sugie, Noboru. 2001. A simple neural network pruning algorithm with application to filter synthesis. *Neural Processing Letters* 13(1): 43–53.

Vapnik, Vladimir, Golowich, Steven E, Smola, Alex et al. 1997. Support vector method for function approximation, regression estimation, and signal processing. *Advances in Neural Information Processing Systems*, 281–287.

# 4

# CONVOLUTIONAL
# NEURAL NETWORKS

In the previous chapters, we studied fully connected multilayer neural networks (MLNNs) and their training using backpropagation. In a typical MLNN layer, with $n$ input nodes and $m$ neurons, we need to learn $n \times m$ parameters or weights. While an MLNN may perform well in some cases, in particular, for those where the features of different dimensions are independent, there are some additional properties in the connection architecture that we might desire. For example, if it is known that the dimensions of the input data are strongly correlated or that the the size of the MLNN (both the number of the layers and the number of neurons in each layer) must be limited for computational considerations, should there be any architectural changes introduced to a standard MLNN to accommodate this additional constraint about the data or the network complexity?

Despite the general expressive power of MLNNs, we might want to make explicit use of local dependencies and invariances among the features themselves. Feature dimensions might have some special ordering of significance, and some change in properties exhibited by neighboring feature dimensions may provide valuable information. In one-dimensional (1D) feature spaces like digital audio signals for voice recognition, or two-dimensional (2D) feature spaces like images, or three-dimensional (3D) feature spaces like videos, and so on, we often need to identify patterns that might occur anywhere in the signal. These patterns may be represented by a template, which may be a short/small-sized signal itself. In all the preceding examples, the data in which we search for the occurrences of a template exhibit strong correlation among nearby samples. Also, from the examples discussed in Chapter 1, many feature representations in images are obtained by convolution with some kernels. These have been among the motivations behind the introduction of convolutional layers to the MLNN (Fukushima, 2003; Fukushima et al., 1983; LeCun et al., 1989). In this chapter, we shall study MLNNs that employ convolutional and pooling layers and their combinations for constructing different deep networks that have been used to produce good performance on many long-standing computer vision problems in recent years.

## Convolution and Pooling Layer

We have already studied sparsity in the context of linear regression before. Sparsity may be defined as the fraction of the zeros in weight matrices or arrays. Having a zero weight or a zero response from a neuron is equivalent to having an incoming or outgoing connection, respectively, removed from the network architecture.

Suppose that the weights were forcibly made sparse for a neuron. The sparsity that is so imposed is random (which is often referred to as dropout and will be discussed later) and the imposition has a structure. This sparsity can be imposed in such a manner that every neuron has incoming weights from only $k$ *adjacent* inputs, adjacency being the important criterion. In this sparse connectivity, we drop out all outputs of a fully connected neuron and only make a few adjacent connections active, making the connectivity sparse and local. For the sake of descriptions, let us consider the adjacency of $k = 3$.

Figure 4.1 depicts such a connection. In this connection setup, each neuron attempts to learn a local property of the inputs as it only takes as input three adjacent features. The property that it tries to learn is perhaps the manner in which the features change across the locality and so forth. *Locality* can be defined in terms of adjacency of dimensionality of signals under a special ordering. A set of connections are local if they are connected to adjacent dimensions in the ordering of the signal. In the case of images, this corresponds to neighboring pixels. One of our aims in looking at locality for images is that we have pixels that are ordered in a sequence and we want to exploit the relationship between pixels in this ordering. There are cases, other than images, where this is also used. For instance, in the case of audio signals the ordering is by time. In the case of images, the ordering is naturally the ordering of pixels in the image itself.

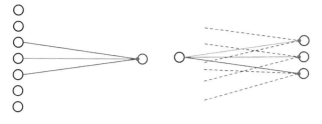

**Figure 4.1**    Sparse and local connectivity.

The connectivity is therefore local and sparse at the same time. Each neuron learns a feature that is local to those it connects to and learns a template that can be used only at those locations. In cases of audio or image signals, this allows us to study features that combine only adjacent local values, such as a word or tone. In images, this allows us to study local features such as gradients, edges, or blobs.

If each neuron has its own local connection and the weights are deemed as some kernel for feature detection, then local connectivity may imply that all features are only local and do not occur elsewhere. This is not true in the case of audio or images. If we are looking for a pattern, say an edge, at some location, we might as well look for the same pattern all throughout the signal. If a neuron is connected to one location in an image where it learns to detect a particular edge pattern, we also want the learned detector to be useful at all possible locations in the image to see if the said edge pattern exists anywhere else. In other words, we want the neuron shown in the left side in Figure 4.1 to not just be connected to those $k$ adjacent input feature locations but slide and circularly shift its connections to other $k$ adjacent input locations as well. In doing so, we do not require new weights for every new connection made, but rather use the same weights at all locations. This effectively allows us to move the neuron around and collect the neuron's response at different locations of the signal. One neuron could go through the entire signal and still only require $k$ weights. Therefore, by doing this, we only increase the number of outputs but not the number of weights.

Note that the outputs are also ordered if arranged properly. If a neuron was an edge detector, it slides around an image trying to find that edge at all locations of the image. The neuron outputs another image that will be *approximately* the same size of the original input image itself.

The output is a locational representation of where the edge is present and where it is absent. This is similar to the edge detector we studied in Chapter 1. In practice, we do not make a neuron move around the signal. Instead, we will have many neurons that share the same weights. While both these are equivalently the same way of thinking about this connection setup, we only consider the second interpretation. By using the second interpretation of weights being shared by neurons at different locations, we are able to retain the original idea of a neuron and its sparse connectivity without thinking about the additional implementation complexity of moving the neurons around.

Let us have a closer look at what this series of operations are accomplishing. Consider that we have a $d$-dimensional feature $x = \left[ x^{(1)}, x^{(2)}, \ldots, x^{(d)} \right]$ as inputs to some layer, which performs this operation. By having a set of weights in some fixed order $w = \left[ w_1, w_2, \ldots, w_k \right]$ and starting from the beginning of the signal and sliding this to the end with a stride of one feature dimension per slide, we are collecting the response of the neuron over each location of the signal given its ordered surroundings. More formally, the operation can be defined as

$$z^{(j)} = \sum_{i=1}^{k} x^{(j+i-1)} w_k, \quad \forall j = [1, 2, \ldots, d - k + 1] \tag{4.1}$$

This operation is often referred to as convolution. To be more accurate, this process is called cross-correlation and not convolution. In the strictest signal processing sense of convolution, the weights $w$ get flipped around its center index. For the sake of convenience and with a slight abuse of notation, we are going to represent both operations using $*$ and refer to both operations interchangeably. While reading the literature though, the reader is well advised to pay attention to this. This is a 1D convolution as opposed to the 2D convolution that was introduced in Chapter 1. The outputs $z^{(j)}$ are the output responses of the neurons that share the same weight. The outputs are ordered responses of the neurons at different locations of the signal. The outputs will be of length $n - k + 1$ as there are that many neurons that share the weights.

These neural responses might also be passed through an elementwise activation function such as a tanh or a rectified linear unit. We call these activated outputs *feature maps* or *activations* interchangeably. The number of feature maps we output is the same as the number of sets of shared-weight neurons or filters (Gardner, 1988). These neurons that share a set of weights are often interchangeably referred to as *filters* or *kernels*. The term *filter* is borrowed from traditional signal processing literature. Figure 4.2 describes this architecture with two filters. Each filter has a set of $k = 3$ weights. Since the filter collects signals from $k$ adjacent input locations at each stride, we say that the filter's receptive field is $k$. A collection of multiple such filters or essentially a layer is called a *filter bank*. The area these filters cover with respect to the original image is often referred to as *receptive fields*.

In Figure 4.2, $k$ is the *receptive field*, as the neuron receives the signal from the original source. The sizes of receptive fields are often debated upon. One school of thought is that filter sizes should be as small as possible, typically on the order of 3 pixels irrespective of the size of the images or signals. This provides an opportunity to learn small-scale features at each layer and build meaningful representations deeper in the network. On the other hand, another school of thought is to learn slightly larger receptive filters at the earlier stages and go smaller at the deeper layers. This debate is not settled, and one needs to design receptive fields suitable to the demands of the application. We will see

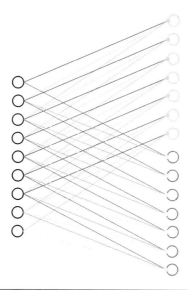

**Figure 4.2** Two sets of shared-weight sparse neurons exhibiting stride 1 convolutional connections. The filter size is 3 and the feature map size (number of neurons sharing the same weights) is 7. The intensity of the lines describes the parameter weights: the darker the color, the lower the weight. Similarly, differently shaded circles represent unique weight sharing between neurons. The signal length is 9 and the filter size is 3, therefore each neural response length is $9 - 3 + 1 = 7$.

this debate over the size of receptive fields develop more quantitatively as we study some case studies later in this chapter.

There are two ways of referring to receptive fields. One way is to think of reception from the previous layer only. In this case, a receptive field of size 3 for any layer is just simply that. Another way to think of receptive fields is to think of them as receiving signals from the input layer. This implies that the receptive field of a second layer of a size 3 filter bank receiving a signal from a first layer of a size 3 filter bank is a size 9. We will follow the former notation here for the sake of convenience, but it is to be noted that deeper layers correspond to a much larger area of reception from the original signal.

A 2D convolutional layer is like a 1D convolutional layer except that the filter weights are also ordered in two dimensions. The first part of Figure 4.3 shows a typical 2D convolutional layer. Each neuron may start from some corner (say, top-left) of the 2D signal (most often images) and stride through them in one direction and end at the opposite corner (say bottom-right). Each input itself might be a collection of feature maps or channels of images. In such a case, we convolve each neuron with every feature map or every channel of the inputs. The outputs of each channel may be added together in a location-wise addition so that each neuron will have one averaged output response. This reduces the number of activations at the output of the layer. Consider that the input to the layer $a$ has $I$ channels (could also be feature maps or activations if coming from another layer), and the

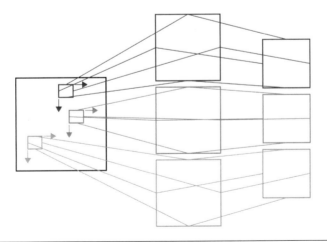

**Figure 4.3**  Convpool layer with three neurons.

layer has $L$ kernels $\boldsymbol{k}$ and $\lambda$ representing the element-wise activation function. The output activations of the layer are

$$\boldsymbol{z}^{j} = \lambda\left(\sum_{i=1}^{I} \boldsymbol{a}^{(i)} * \boldsymbol{k}_{j}\right) \forall \; j = [1, 2, \ldots, L] \qquad (4.2)$$

The * represents the convolution operation that we discussed in Chapter 1. The difference being that instead of handcrafting the filters as we did in Chapter 1, here the filters are learned by the network itself. The filters learned by convolutional layers that directly work on the input images are often similar to edge or blob detectors. These represent some properties or types of changes locally in the input image. Figure 4.4 shows some examples of filters that were learned from the CIFAR-10 dataset. It also shows the output activations of one of the filters (Krizhevsky and Hinton, 2009).

Neural networks are not mere learning systems that learn concepts but also systems that break down data and represent (explain) data using smaller pieces of information. Using these representations, the neural network maps data to the label space. Convolutional neural networks (CNNs), for instance, break the image into small basic components of images. These are typically edges and blobs, which we would find as local properties commonly among images in a dataset. By using these edges and blobs in the right manner at the right locations and adding shifted versions, we may sometimes be able to recreate the original image itself. We shall study this in detail as an autoencoder in

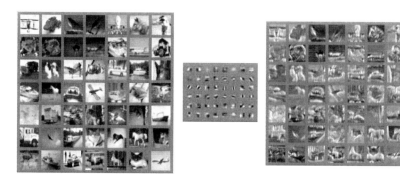

**Figure 4.4**  From left to right are some images from the CIFAR-10 dataset, some 5×5 2-D convolutional filters that were learned from it, and the activations that the top-left filter produced. The code that is used to learn this network is available in the Yann toolbox (refer to Appendix A).

Chapter 5. Others use these techniques to map the images onto more complex spaces than the simple label space that we have been studying thus far. Consider semantic segmentation, object localization, or generating images of a higher resolution. In these cases, on top of the network predictions (which are already images), we may add other structured models such as Markovian Random Fields and produce structured predictions. In this chapter, we continue using only simple label spaces. After successive transformations and retransformations with projections on better feature spaces, the network maps the image onto a discriminative label space.

In identifying what kind of an object is present in the image, the CNN will break the image into a representation on what edge or blob is present at which location. This effect is easily observable in the first layer. Second-layer weights are trying to find representations on the locations of first layers' blob and edge activations. Deeper layers have more complicated representations, which often do not have a human-interpretable meaning.

In a convolutional layer, the neuron activations of nearby locations often represent similar features. Consider a filter that is a pattern detector for a particular *edge*, a filter perhaps of the form $[-1, 0, 1]$. As we already saw in Chapter 1, this filter produces an output for every point where there is a horizontal gradient. We may use the activation function as a thresholding function perhaps, but we might still end up with more than one-pixel-thick edge segments that do not define the edge nicely.

The activations change smoothly since the convolutional filter is small and the range or pixels looked at is large. Sharper transitioning activations make for better learning. To achieve this, a *pooling* or a subsampling layer typically follows the convolution layer although this need not always be true. This can be achieved also by using a strided convolution layer or by using a combination of strided convolutions and pooling. Pooling is generally considered one way of achieving this, although recent studies use strided convolutions to replace pooling altogether (Radford et al., 2015; Visin et al., 2015). A strided convolution is a convolution when instead of moving the mask by one step at a time, we stride by a larger step.

By pooling, we are in general reducing the data entropy by reducing the size of the activations. Reduction of size is often good because, while we lose some spatial information and spatial frequency in an activation itself, we gain a lot more activation responses through a layer's filter bank. More importantly, pooling allows us to impose and accommodate for some invariances among features spatially. The deeper we go in a neural network, the more activations we will get the and larger will be the sizes of these activations. This will soon become computationally intractable. Pooling helps us in maintaining tractability.

The norm in subsampling from a signal processing perspective is to pool by average, but maximum pooling or *maxpool* is generally preferred in CNNs. To perform a maxpool by $p$, we choose a sliding window of $p \times p$. Similar to convolution, this sliding window can be strided or of stride 1. Typically, pooling is performed with a stride the same as the size of the pooling widow itself $(p \times p)$. Once within a window, we select the maximum (or the type of pooling) and represent the entirety of that window by that value. Maxpool is more popular than others as it represents the strongest response. In image processing contexts, the strongest response corresponds to the best match of the template the filter is looking for and is therefore a good option. There are some concerns about pooling as it loses information and some attempts have been made to solve it, which we shall see in Chapter 5.

Figure 4.3 shows a typical stacked convolution pooling or *convpool* layer. In a convpool layer, the weights are few and there are a lot of feature maps each representing its own feature. Typically, the filter sizes are small but are still larger than the strides. Pooling window size is typically smaller or the same as the convolutional filter size. Convpool layers are typically implemented as one layer, although they may be refered to in literature as different layers as well.

### Convolutional Neural Networks

A CNN is a neural network where a signal feeds into a set of stacked convolutional pooling layer pairs (convpool layers), and the output of the last layer feeds into a set of stacked fully connected layers that feed into a softmax layer (Figure 4.5).

A similar setup can be created for autoencoders and other neural networks but as we have done in the chapters before, we shall stick to only classification networks and revisit other architectures in Chapter 5. A fully connected neural network as seen in the previous chapters might or might not follow the convolutional layers before

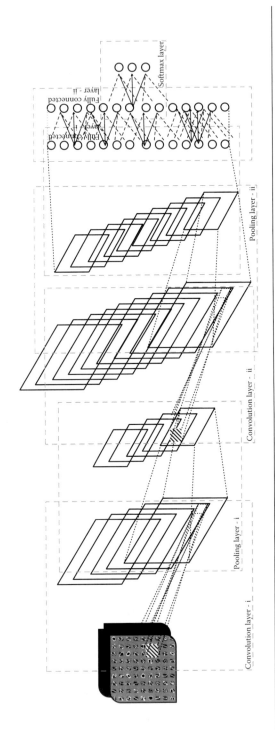

**Figure 4.5**  A complete convolutional neural network. (LeNet)

we go into a typical softmax layer for classification purposes. Since the convpool layer is differentiable, learning in a CNN is the same as learning in an MLNN, which we saw in Chapter 3. Other optimization procedures we studied work for CNNs as well.

While the basic idea of building a CNN is clear from the above presentation, in practice, a designer may choose to configure the network in different ways, resulting in different architectures that are often motivated by problem-specific considerations. In the following, we examine several architectures that have been relatively well known, and in doing so, we hope to illustrate some typical ways of building a CNN of a particular philosophy.

## CASE STUDIES

Now that we have constructed a CNN, let us study some popular CNN architectures and CNN-based applications.

## CASE STUDY 0: THE MNIST DATASET

Before we get into our first network architecture, let us study one of the most commonly used toy datasets for computer vision: the MNIST dataset of handwritten characters, more commonly just MNIST. Knowing this dataset will give us some context with which to study our first CNN architecture case because it was designed for this very application using this dataset.

MNIST is a grayscale image dataset that was created as a replacement for the NIST dataset, which was black and white. MNIST contains tightly cropped images of handwritten numerical characters. Each image contains $28 \times 28$ pixels with one character in its center. The dataset has 70,000 images roughly evenly distributed among all the classes. A total of 50,000 of these images are used for training, 10,000 for testing, and 10,000 for validation. These splits are premade and are constantly maintained. Figure 4.6 shows some sample images from this dataset.

The constancy of the image sizes, scales, and writing makes this dataset an excellent tool to study and test machine learning algorithms at a preliminary stage. Since the dataset was also well arranged, while working with CNNs we can notice the actions

**Figure 4.6**   Sample images from the MNIST dataset.

of each filter that we learn, making this a perfect toy dataset to study CNNs with. As is tradition with CNNs, we shall begin the study of CNNs with the MNIST dataset.

### CASE STUDY 1: LENET

One of the earliest and one of the most popular CNNs is Prof. Yann LeCun's CNN for digit recognition (LeCun et al., 1998). This network is now often referred to as LeNet5, or simply LeNet. LeNet5 is a CNN with two convolutional layers and one fully connected layer, whose detailed network architecture is shown in Table 4.1.

The LeNet5 for MNIST takes in the $28 \times 28$ single-channel grayscale images as input to the first layer. The first layer has a receptive field of $5 \times 5$. Originally, it only had six feature maps for computational reasons, but the widely used modern incarnation of this network has 20 feature maps in the first

**Table 4.1**   A Modern-Day Reincarnation of LeNet for MNIST Classification

| LAYER NUMBER | INPUT SHAPE | RECEPTIVE FIELD | NUMBER OF FEATURE MAPS | TYPE OF NEURON |
|---|---|---|---|---|
| 1 | $28 \times 28 \times 1$ | $5 \times 5$ | 20 | Convolutional |
| 2 | $24 \times 24 \times 20$ | $2 \times 2$ | – | Pooling |
| 3 | $12 \times 12 \times 20$ | $5 \times 5$ | 50 | Convolutional |
| 4 | $8 \times 8 \times 50$ | $2 \times 2$ | – | Pooling |
| 5 | 800 | – | 500 | Fully connected |
| 6 | 500 | – | 10 | Softmax |

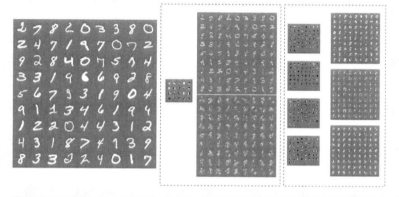

**Figure 4.7**   Filters and activations of the LeNet5 convpool layers on MNIST images.

convolutional layer. The output of this layer has 20 activations of size $28 - 5 + 1 = 24$ ($\times 24$). This gets pooled down by a factor of 2, making an input to the next convpool layer of $12 \times 12$. The next layer learns 50 feature maps, producing 50 activations of $8 \times 8$ each. This gets downsampled to $4 \times 4$. Before going into the fully connected layers, we *flatten* these activations into $1 \times 800$, which feed into a typical dot-product hidden layer of 500 neurons. This then goes to a softmax layer of 10 nodes, one for each character in the MNIST dataset. In today's LeNet5, all activations are ReLU units providing for faster learning. Figure 4.7 shows some the filters learned and the activations of some of the learned filters for the convolutional layers of LeNet5. This is after training the network for 75 epochs with a learning rate of 0.01. The CNN of this nature, trained so, produces an accuracy of 99.38% on the MNIST dataset. The code for producing these results and figures are available at the yann toolbox's tutorial, refer to Appendix A for more details.

Now that we have seen the MNIST dataset and the LeNet5, which is a prototypical CNN, we will proceed to a few modern-day CNNs that were influential in the resurgence of CNN-based techniques for computer vision problems. This resurgence was due to a variety of factors including advancement in computing technologies (powerful CPUs but more importantly the availability of GPUs supporting parallel computing), larger and better image datasets, open visual object categorization (VOC), ImageNet competitions, and others (Russakovsky et al., 2015). Each of the networks studied here represents some significant progresses made at the time and hence they serve very well to illustrate how new ideas may be incorporated to improve upon the basic CNN architecture.

## CASE STUDY 2: ALEXNET

Krizhevsky et al. introduced their CNN at the Neural Information Processing Systems (NIPS) 2012 conference (Krizhevsky et al., 2012), which is often referred to as AlexNet. This network was the winner of the ImageNet challenge of 2012 (Russakovsky et al., 2015).

Similar to the MNIST being influential in making the LeNet5 widely known and used, ImageNet is one large-scale dataset that was heavily relied upon by many researchers in the *third wave* of neural networks. At the time of AlexNet, ImageNet was a dataset that had over 15 million images in over 22,000 categories. The ImageNet competition, ImageNet large-scale visual recognition challenge (ILSVRC), used about 1000 of these categories. ImageNet categories are much more complicated than other datasets and are often difficult even for humans to categorize perfectly. The average human-level performance is about 96% on this dataset. Recognition systems using conventional techniques, such as those discussed in Chapter 1, may achieve at best a performance with only around 75% accuracy, and in 2012 AlexNet was the first system to break the 80% mark (Russakovsky et al., 2015).

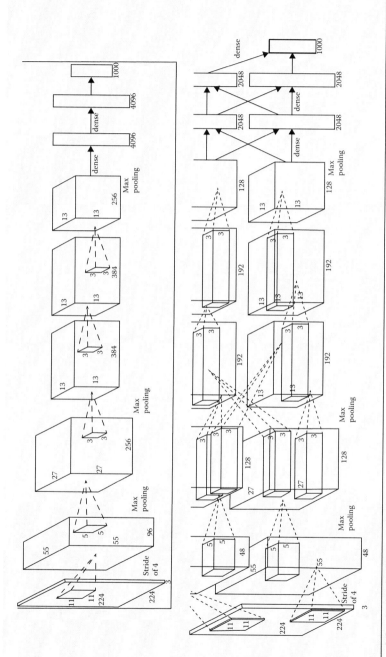

**Figure 4.8**     AlexNet architecture. (Courtesy of Alex Krizhevsky.)

AlexNet at its core is just simply a deeper LeNet. The top network in Figure 4.8 shows the entire architecture of AlexNet. It has five convpool layers followed by two fully connected layers and a 1000-node softmax layer, one for each category. Although at first glance this network looks like a straightforward extension and deepening of LeNet in terms of its architectural philosophy, the sheer number of parameters and therefore the complexity of the model leads to drastic amounts of overfitting. This network by itself could never be trained in a stable manner with the ImageNet dataset even though there are over 15 million images. Computationally, training such a large network is also a major concern. At the time of its implementation, AlexNet was one of the largest and deepest neural networks. Besides the basic training algorithm for this CNN using stochastic gradient descent, Krizhevsky et al. had to use several new techniques in combination to get this network to train without overfitting and in a fast-enough manner. We will study these techniques in detail below.

One significant novelty of AlexNet was the choice to use the ReLU activation instead of the then more traditional tanh or sigmoidal activations (Nair and Hinton, 2010). Tanh and sigmoid activations, as we discussed in the previous chapters, are *saturating activations*. ReLUs do not saturate near 1 like other activation functions and therefore still have a larger gradient as we approach unity. This helps to increase the speed of learning.

Another important contribution of AlexNet is in its use of multiple GPUs for learning. The original implementation is still available as *cuda-convnet*, although advanced versions have been implemented since then in the form of many toolboxes. The bottom image in Figure 4.8 shows the network being split into two. This is because one part of the network runs on one GPU while the other runs on another. Krizhevsky et al. originally trained the networks on two Nvidia GTX 580 GPUs with 3 GB worth of memory and were not able to fit the entire network in one GPU. This split makes the network learn very specialized features.

Figure 4.9 (top) shows filters learned in the training process. The top three rows are typically *color-agnostic* features that are learned on one of the GPUs and the last three rows are

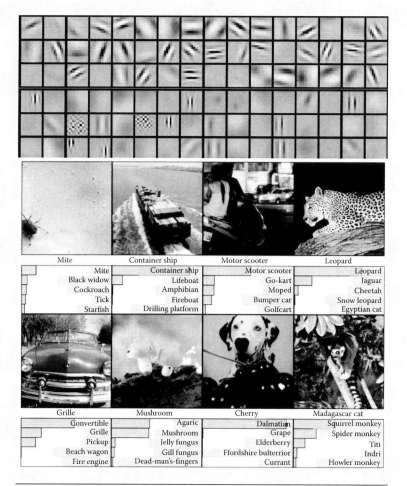

**Figure 4.9**  AlexNet filters learned at the first layer and sample results and predictions. (Courtesy of Alex Kirzhevsky.) The production of this figure is in grayscale. Color productions of this image could be found in the original article (Krizhevsky et. al., 2012).

*color-sensitive* features learned on the other GPU. Such a behavior was seen to exhibit over many repetitions. Even though it was trained in two parallel GPUs, they do communicate and are not completely independent multicolumn networks.

Yet another significant advancement in AlexNet was the use of dropouts. AlexNet is a network with a large number of parameters and training a network of this size without proper regularization will lead to severe overfitting. Dropout is a method to avoid coadaptation of feature maps (Srivastava et al., 2014). Ideally, to improve performance, we would like to train many models and combine their decisions by taking a poll average of all the models we have trained. This technique is called ensemble learning and is typically used in random forests and decision trees. When it comes to neural networks, ensemble learning poses a computational problem. Training one AlexNet was already difficult on two GPUs; to train tens or even hundreds of them would be a logistical impossibility. Dropout is one method that addresses these problems. Dropout is a technique where the output of a neuron is forced to zero with a random probability. This random probability is typically 0.5 Bernoulli. This means that for every neuron, we associate a random variable from which we draw a state from a Bernoulli distribution with a probability of 0.5. If the sample drawn was 1, we allow the signal to go through the neuron; if it was 0, we drop the output of the neuron to 0, irrespective of whether the neuron was active or not. This implies that approximately half the neurons in the network typically turn off randomly while learning. This creates many advantageous effects.

Since the network is now running with only half the representations, each neuron is now under twice the *stress* to learn meaningful features. This enables faster learning without overfitting. Since neurons turn OFF and ON at random, neurons can no longer coadapt and must learn features that are independent of other neurons. More importantly, during each backpropagation effectively a new architecture is created by using dropouts. Each forward propagation is a whole new network model that we are learning in our ensemble. The one difference is that these models now share parameters. Dropouts are not typically used in a convolutional layer in the traditional context as they would remove an entire feature map. Dropouts are applied to individual feature responses dropping out only a few locations in the feature maps randomly, which in itself isn't ideal.

Table 4.2 shows the complete AlexNet architecture in detail. The results produced by AlexNet were extraordinary for the time and yet this architecture was limited only by computational and memory capacities. The authors predicted better accuracies

**Table 4.2**   AlexNet Architecture

| LAYER NUMBER | INPUT SHAPE | RECEPTIVE FIELD | NUMBER OF KERNELS | TYPE OF NEURONS |
|---|---|---|---|---|
| 1 | 224 × 224 × 3 | 11 × 11, stride 4 | 96 | Convolutional |
| 2 | – | 3 × 3, stride 2 | – | Pooling |
| 3 | 55 × 55 × 96 | 5 × 5 | 256 | Convolutional |
| 4 | – | 3 × 3, stride 2 | – | Pooling |
| 5 | 13 × 13 × 256 | 3 × 3, padded | 384 | Convolutional |
| 6 | 13 × 13 × 384 | 3 × 3, padded | 384 | Convolutional |
| 7 | 13 × 13 × 384 | 3 × 3 | 256 | Convolutional |
| 8 | 30,976 | – | 4096 | Fully connected |
| 9 | 4096 | – | 4096 | Fully connected |
| 10 | 4096 | – | 1000 | Softmax |

were to come with deeper and more complicated models. This prophecy has come true many times over and led to some other more recent networks that we shall see in the rest of this chapter and in Chapter 5.

Beyond the objective results in terms of winning the competition and producing a significant improvement in accuracy on the ImageNet dataset, AlexNet also demonstrated some interesting subjective results, as illustrated in Figure 4.9. Notice that, for instance, not only are the first choices of labels good, but even the other labels predicted by the system were mostly quite semantically related (and in some cases corrected the labeling error of the dataset itself). For instance, the additional guesses for leopard are snow leopard and jaguar, which are not such bad alternative options. Such subjective results appear to demonstrate that AlexNet may have learned to represent the underlying objects in a more semantic manner that groups semantically related objects together. We will further investigate this result in Chapter 5.

## CASE STUDY 3: GOOGLENET AND THE INCEPTION MODULE

In 2015, Google came up with a special convolutional layer called the inception layer (Szegedy et al., 2015). There is a fundamental shortcoming with an AlexNet type of network: large computational cost while at the same time providing depth. The inception module tries to make a network with fewer parameters and at the same time go deeper. The inception module is a network-in-network system. Figure 4.10 shows an inception module. Each inception module contains a few convolution layers in parallel that go through a dimensionality reduction step through $1 \times 1$ convolutional layers. There is one maxpooling layer. These layers get concatenated before being passed on to another inception or a regular module. The $1 \times 1$ convolutions reduce the dimensionality similar to a technique called embedding. This gets passed on to the more expensive $5 \times 5$ and $3 \times 3$ layers. The embedding layers perform a version of clustering before being passed on the next layer. Capsules are a similar and related idea, which takes

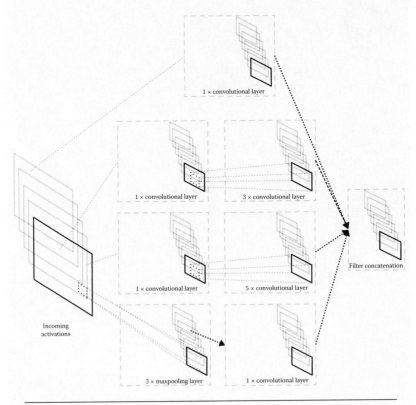

**Figure 4.10**   GoogLeNet's inception module.

this further in doing an actual Hough transform like voting to do clustering (Hinton et al., 2011).

*GoogLeNet* is one particular network that uses the inception module. Not including pooling layers, this network contains 22 layers. With such a deep network, there arises a problem of vanishing gradients. Vanishing gradient is a problem of depth wherein the errors are not strong enough to produce gradients that are strong enough to move the weights in any direction. To avoid this, in between, layers are branched off into one fully connected softmax layer. The point of these classifier layers is not to perform better in label accuracies but to add more errors so that there are some gradients that produce additional discriminative features. These act as a regularizer. By adding additional softmax layers in between the network, we can create errors at those points that help in producing more supervision, thereby stronger gradients. Effectively, we are regularizing the network to be always discriminative at all layers.

One quirk of GoogLeNet is having traditional convolution layers at the beginning followed by inception modules. The traditional convolution layers produce activations that can be easily clustered upon using the 1 × 1 layers. The fully connected layers were trained with a 0.7 dropout unlike the 0.5 of AlexNet and a 0.9 momentum. The results of this network are not reported here and are left for the interested reader to look up from the original reference. GoogLeNet was able to significantly increase the state of the art in several datasets including ImageNet.

## CASE STUDY 4: VGG-19

The VGG network is a popular benchmark network that beats the GoogLeNet postcompetition on the ImageNet 2014 (Simonyan and Zisserman, 2014). The VGG network works on a filter size philosophy different from the ones we discussed in the previous networks. In all the previous networks, we considered filters that are larger in size at the earlier layers and smaller in size at the deeper layers. VGG follows the same idea but starts with a receptive field that is as small as 3 × 3 and grows at a constant pace of

$3 \times 3$. The VGG network considers only $3 \times 3$ convolutional filters with a $2 \times 2$ maxpooling layer and stacks them to create a convolutional network. This network is very close to the LeNet in all its architectural paradigms, with no inception modules or the like. The original VGG paper enlists five different architectures, but we are interested in only the fifth (E) as it is the most commonly used one. The architecture of interest is shown in Table 4.3. The E network is a simple convolutional network that is 24 layers deep and is commonly referred to as the VGGNet or VGG-19.

The strategy to build up a large network layerwise using same-sized smaller filters was originally to analyze the effect of depth

**Table 4.3**   VGG Network

| LAYER NUMBER | RECEPTIVE FIELD | NUMBER OF KERNELS | TYPE OF NEURONS |
|---|---|---|---|
| 1 | $3 \times 3$, stride 1 | 64 | Convolutional |
| 2 | $3 \times 3$, stride 1 | 64 | Convolutional |
| 3 | $2 \times 2$, stride 1 | – | Pooling |
| 4 | $3 \times 3$, stride 1 | 128 | Convolutional |
| 5 | $3 \times 3$, stride 1 | 128 | Convolutional |
| 6 | $2 \times 2$, stride 1 | – | Pooling |
| 7 | $3 \times 3$, stride 1 | 256 | Convolutional |
| 8 | $3 \times 3$, stride 1 | 256 | Convolutional |
| 9 | $3 \times 3$, stride 1 | 256 | Convolutional |
| 10 | $3 \times 3$, stride 1 | 256 | Convolutional |
| 11 | $2 \times 2$ stride 1 | – | Pooling |
| 12 | $3 \times 3$, stride 1 | 512 | Convolutional |
| 13 | $3 \times 3$, stride 1 | 512 | Convolutional |
| 14 | $3 \times 3$, stride 1 | 512 | Convolutional |
| 15 | $3 \times 3$, stride 1 | 512 | Convolutional |
| 16 | $2 \times 2$, stride 1 | – | Pooling |
| 17 | $3 \times 3$, stride 1 | 512 | Convolutional |
| 18 | $3 \times 3$, stride 1 | 512 | Convolutional |
| 19 | $3 \times 3$, stride 1 | 512 | Convolutional |
| 20 | $3 \times 3$, stride 1 | 512 | Convolutional |
| 21 | $2 \times 2$, stride 1 | – | Pooling |
| 22 | – | 4096 | Fully connected |
| 23 | – | 4096 | Fully connected |
| 24 | – | 1000 | Softmax |

in performance and as was expected the study finds that the deeper the network, the better the performance. The claim of deeper being better has been challenged by some recent results and is therefore not a settled conclusion, although the results reported by VGG work for layers of the sizes studied. Being a 3 × 3 filter on the first layer, it is interesting to see the kind of templates the network is learning to detect. Figure 4.11 shows all the first-layer filters of the VGG19 network. Notice that the filters are predominantly corner detectors, edge detectors, and color blobs. These are the fundamental building blocks to build higher order features as we discussed in previous chapters. VGG19 has become a prototypical network and its first-layer features appear to be further evidence that CNNs can be used to learn feature detectors, similar to those we studied in Chapter 1.

In the original work, to train the VGG network, a simple 0.9 momentum was used with a change in learning rate once the performance saturated. The training was completed in 74 epochs with a batch size of 256. With this network, VGG could just beat

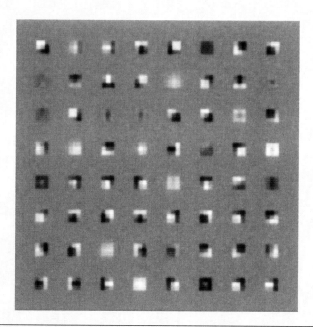

**Figure 4.11**  First-layer features of the VGG-19 network. Refer to the book website (convolution.network) for a color rendering of this image.

GoogLeNet. Note that VGG is shallower than the GoogLeNet, leading many to stipulate that smaller and consistent filter sizes is a good if not a better technique, although the debate is still open on this question as well.

## CASE STUDY 5: RESIDUAL NET

In GoogLeNet, we found an architecture for a layer, where the layer itself separates into a few branches and produces a summed or concatenated output. A layer that ends up in a summation or concatenation splits the gradient into equal parts and passes the gradient through during backpropagation. Proposed by He et al., residual networks or ResNets are a novel architecture, which uses the idea of splitting a layer into two branches, where one branch does nothing to the signal and the other processes it as would a typical layer (He et al., 2015). The unprocessed data or the residual is added to the original signal going through the network unaltered. This creates a split in the network, where one branch quite simply propagates the gradient through without altering it. This lets a deeper network learn with strong gradients passing through. While in both GoogLeNet and in ResNet the data (and therefore the gradient) passes through, the inception module still had a pooling layer, which is avoided completely by the ResNet module. Figure 4.12 illustrates this idea. Note that there are some articles that demonstrate that a very deep ResNet is equivalent to a shallow recurrent neural network (where a layer feeds on its temporally past self, with or without a store memory sate) (Liao and Poggio, 2016) and a similar architecture called highway nets was also proposed, which was inspired from recurrent neural networks (Srivastava et al., 2015).

Several modifications have been proposed to this original ResNet architecture. One idea is to create an architecture where some layers might be turned off randomly during training time and compensated for during the test time quite akin to dropouts (Huang et al., 2016). This network architecture will replace the dropped module using an identity (implying that the processed part of the residual simply produces a zero). While it is counterintuitive to drop layers at training time, one must note that the

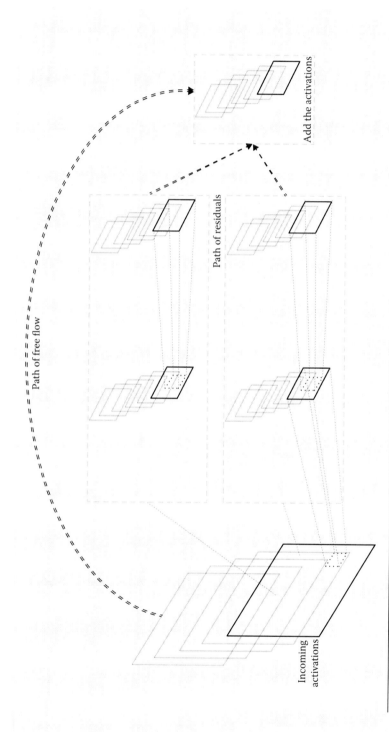

**Figure 4.12** A typical residual layer architecture. There are several (one) modules that are parameterized that are accompanied by a connection where the gradient shall flow unaffected. The final output of the layer is the sum of all the activations produced with the original input to the layer.

network at test time is still a deep network and at training time has stochastic depth (the depth is random). One result of the ResNet architecture is that, while traditionally arbitrarily deep networks tend to overfit and sometimes even produce poorer performance than a network with fewer layers, ResNets seem to hold true to the idea that increased depth of the network implies increased performance. ResNets beat the performance records set by all the previously discussed networks and one of its variants holds the record on the ImageNet challenge at the time of writing this book. The ResNet that won the ImageNet competition has up to 152 layers and also has an implementation at the time of 1,000 layers for the CIFAR-10 dataset.

## Summary

In this chapter, we studied sparse and local connections instead of the complete mesh connection from Chapter 3. We also introduced weight sharing as a method to reduce the number of parameters in a network. Not only was it a way to reduce the number of parameters but it was also a way to detect the same feature across the façade of the image. Sharing weights helps us identify similar and recurring patterns throughout the image. Convolution layers are one way to achieve both weight sharing and local and sparse connectivity. We also identified pooling as a way to reduce the size of the activations produced by convolution. This is important as convolution layers produce activations that are memory inefficient. Pooling also helps us in using features that are spatially invariant.

We made use of convolutional and pooling layers to create the convpool layers. The convpool layers are the workhorse of modern-day CNNs for neural computer vision. We also studied various properties of the convpool layer. We noticed that the early layers of the CNN using convpool layers learn filters that resemble edge detectors or blob detectors. We also studied several popular CNNs such as LeNet, AlexNet, GoogLeNet (InceptionNet), and VGG-19. Along the way, we studied some special layers or techniques that these networks employ including dropouts, the question of filter sizes, and the inception module. We also studied the ResNet architecture and noticed that free-flowing gradients provided for stable learning of deeper networks.

It is interesting to note that there have been studies that seem to suggest that there may be a process in the primate visual cortex similar to the convolutional processing in a CNN (Hubel and Wiesel, 1968), especially as far as the concept of receptive field goes, although we caution a reader about such interpretations since there are many "tricks" (such as training with dropout) in a CNN that probably have only computational meanings, but not necessarily any biological correspondence.

# References

Fukushima, Kunihiko. 2003. Neocognitron for handwritten digit recognition. *Neurocomputing* 51: 161–180.

Fukushima, Kunihiko, Miyake, Sei, and Ito, Takayuki. 1983. Neocognitron: A neural network model for a mechanism of visual pattern recognition. *IEEE Transactions on Systems, Man, and Cybernetics* 13(5): 826–834.

Gardner, Elizabeth. 1988. The space of interactions in neural network models. *Journal of Physics A: Mathematical and General* 21: 257.

He, Kaiming, Zhang, Xiangyu, Ren, Shaoqing et al. 2015. *Deep residual learning for image recognition.* arXiv preprint arXiv:1512.03385.

Hinton, Geoffrey E, Krizhevsky, Alex, and Wang, Sida D. 2011. Transforming auto-encoders. *International Conference on Artificial Neural Networks*, Springer, pp. 44–51, Espoo, Finland.

Huang, Gao, Sun, Yu, Liu, Zhuang et al. 2016. *Deep networks with stochastic depth.* arXiv preprint arXiv:1603.09382.

Hubel, David H and Wiesel, Torsten N. 1968. Receptive fields and functional architecture of monkey striate cortex. *Journal of Physiology* 195: 215–243.

Krizhevsky, Alex and Hinton, Geoffrey. 2009. *Learning multiple layers of features from tiny images*: Citeseer.

Krizhevsky, Alex, Sutskever, Ilya, and Geoffrey, Hinton E. 2012. ImageNet classification with deep convolutional neural networks. *Advances in Neural Information Processing Systems*: 1097–1105.

LeCun, Yann, Boser, Bernhard, Denker, John S et al. 1989. Backpropagation applied to handwritten zip code recognition. *Neural Computation* 1: 541–551.

LeCun, Yann, Bottou, Leon, Bengio, Yoshua et al. 1998. Gradient-based learning applied to document recognition. *Proceedings of the IEEE* 86: 2278–2324.

Liao, Qianli and Poggio, Tomaso. 2016. *Bridging the gaps between residual learning, recurrent neural networks and visual cortex.* arXiv preprint arXiv:1604.03640.

Nair, Vinod and Hinton, Geoffery. 2010. Rectified linear units improve restricted Boltzmann machines. *International Conference on Machine Learning*, Haifa, Israel, pp. 807–814.

Radford, Alec, Metz, Luke, and Chintala, Soumith. 2015. *Unsupervised representation learning with deep convolutional generative adversarial networks.* arXiv preprint arXiv:1511.06434.

Russakovsky, Olga, Deng, Jia, Su, Hao et al. 2015. ImageNet large scale visual recognition challenge. *International Journal of Computer Vision* 115: 211–252.

Simonyan, Karen and Zisserman, Andrew. 2014. *Very deep convolutional networks for large-scale image recognition.* arXiv preprint arXiv:1409.1556.

Srivastava, Nitish, Hinton, Geoffrey E, Krizhevsky, Alex et al. 2014. Dropout: A simple way to prevent neural networks from overfitting. *Journal of Machine Learning Research* 15: 1929–1958.

Srivastava, Rupesh K, Greff, Klaus, and Schmidhuber, Jürgen. 2015. Training very deep networks. *Advances in Neural Information Processing Systems, 2377-2385.*

Szegedy, Christian, Liu, Wei, Jia, Yangqing et al. 2015. Going deeper with convolutions. *Proceedings of the IEEE, Conference on Computer Vision and Pattern Recognition*, pp. 1–9, Boston, MA.

Visin, Francesco, Kastner, Kyle, Cho, Kyunghyun et al. 2015. *ReNet: A recurrent neural network based alternative to convolutional networks.* arXiv preprint arXiv:1505.00393.

# 5

# MODERN AND NOVEL
# USAGES OF CNNS

In Chapter 4, we studied some traditional and some modern convolutional neural networks (CNNs) for image categorization. These are often referred to as *vanilla* CNNs as they are the most straightforward implementations of these networks. The convpool layer and the neural architecture are versatile and can be used for much more than simple image categorization problems. Thus far, we have predominantly studied only supervised learning. CNNs are also good at solving various other problems and have been an active area of research. CNNs, including many of their variants, with additional components of other types of network architectures, are being used in many novel ways for a plethora of applications including, object localization, scene detection, segmentation among others. While these are important applications, in this chapter, we restrict ourselves to some interesting applications of the convpool layer and the CNN architecture for computer vision applications.

In Chapter 4, we built several classifier networks and noticed that the earlier layers of CNNs extract representations that were detectors for basic building blocks of images, such as edges and blobs. We also noticed that several datasets could lead to similar kinds of features on these layers. The similarity in these representations begs the question, *Could we learn a network from one dataset and use the learned representation to classify another dataset?* The simple answer is that we could. Several networks that we saw in Chapter 4 such as ResNet, VGG-19, GoogLenet, and AlexNet all have their weights and architecture in the public domain making them pretrained, downloadable, and off-the-shelf. Training large networks is a sophisticated and often a difficult engineering task. Networks crash due to weight explosion, vanishing gradients, improper regularizations, and lack of memory or compute power. Choosing hyperparameters such as learning rates is also notoriously difficult. The deep learning community itself is very helpful in publishing these pretrained networks and allowing others to borrow and use them. This implies that anyone building a deep network intending to train on a small dataset can instead use these pretrained networks as initializations and *fine-tune* them for his or her task. Fine-tuning a pretrained network has been a common practice in the neural computer vision community. Mostly, these networks were pretrained using the ImageNet dataset. Will it matter therefore if we were to use the representations from these *off-the-shelf* networks as initializations to fine-tune on a different dataset even if the new dataset potentially has statistics in contrast to the ImageNet dataset? The following section will survey this very problem.

### Pretrained Networks

Image datasets have constantly grown in sophistication. The MNIST-like datasets of the previous decade were structured and controlled to such a degree that they were not *natural*. Contemporary datasets such as ImageNet, PASCAL, or Caltech-101/256 contain images like what one might click on one's personal camera (Everingham, n.d.; Fei-Fei et al., 2006; Russakovsky et al., 2015). These visual object classification (VOC) datasets are what we now refer to as *naturalized image datasets*—images that are not from a heavily controlled setting but are direct photographs sampled from the real world. Since all these datasets contain natural images, the images in them ought to have similar statistical properties and therefore might lead to comparable feature representations on a CNN trained using them.

Some datasets have more images and/or number of classes than others. Therefore, some datasets provide more diversity of images to learn features than others. Since the images across these datasets also appear similar, an argument could be made that a network trained on one dataset could be used as a feature extractor for another without explicitly being learned or simply fine-tuned from the latter. Suppose that we have in our possession two datasets, $A$ and $B$, where $A$ has considerably more images and object classes than $B$. If we were to build a CNN to learn to categorize only $B$, and we use the training set of $B$ alone, there is a chance that we might overfit. Also, with fewer images, it is difficult to train deeper and larger CNNs. If we were to train a network on $A$ first and destroy the softmax layer, we could now rebuild another softmax layer for the categorization of $B$. All the layers but the last softmax layer are carried over in the hope that two datasets share the same features.

To understand how this trick works, let us consider only the first layer of a CNN. In Chapter 4, we noticed that the first layer learns edge detectors and blob detectors. In Chapter 1, we also saw how some detectors such as the histogram of oriented gradients (HOG) could be used as feature detectors for most datasets if we are interested in using shape for categorization. Once we have a CNN trained on a dataset, we can think of the network as containing two distinct components: a feature extractor and a classifier.

Figure 5.1 shows a fully trained CNN. The CNN until its last layer can be considered as a feature extractor. The output activation of the last layer is the feature vector and the softmax layer is the classifier that we learn for categorization. In this perspective, we can now think of the entire network but the last layer as one *feature extractor* quite akin to the HOG. The only difference is that instead of handcrafting features, we have learned what is supposedly a *good* feature for that dataset using the images in the dataset itself, or using another (larger) dataset. The feature vector from the last-but-classifier layer is a representation of the image, just as how we created representations of employees in Chapter 2.

The last layer is now nothing but a simple logistic regression as seen in Chapter 3. In actuality, this is not the case because we cannot separate the network from the classifier completely. Coadaptation plays a significant role in these networks being learned, so learning

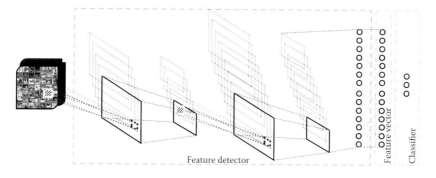

**Figure 5.1**  A CNN as a feature detector followed by a classifier.

the weights along with the classifier is just as important. For the sake of this analysis, this view of the network works in our favor.

In this perspective, one can easily argue that features learned from one dataset could be carried over to another if the images in the datasets *appear* similar and hold similar properties (e.g., local image properties and structures). A more fundamental argument is that most of the fundamental building blocks for many image datasets are nearly the same. The intuition behind this perspective of the network being a universal feature extractor is that the discriminative characteristics that are learned from one dataset are more or less what we expect in other datasets as well. In reality, this argument gets weaker and weaker, the deeper we go into a network as deeper layers are usually more object and dataset specific. This is similar to the argument for creating generalized features such as HOG in the first place. HOGs were expected to produce feature spaces that were discriminative of the shape of the entities in the images, irrespective of what the image dataset was and what the categorization task was. If HOG which was handmade and was general enough to work with many datasets and performed well on ImageNet, a network trained on ImageNet that performs well on it should arguably be general enough to work on other datasets too.

To further ensure that the features are indeed well suited, we need not directly use these as standard feature extractors. We can simply initialize a new network with the feature extractor part of another network trained on a larger dataset and retrain or *fine-tune* the new dataset starting from the weights as they were after being trained on the old dataset. This idea of using pretrained networks to learn categorization on more *specialized* dataset, trained from networks learned from a general dataset is a very strong approach to avoid overfitting and faster training. It could help to avoid overfitting if the initializing network was well trained on a previous (large) dataset, trained and it provides faster training because we are expected to already be close to some good local minima at the beginning. This is akin to a one-time regularization step.

The purpose of the feature extractor is to map the images to a space that is discriminative. Many such networks have been trained using large datasets such as ImageNet and have been made *publicly downloadable* for fine-tuning on a specialized dataset (Russakovsky et al., 2015). The networks that we discussed in previous chapters—AlexNet, GoogLeNet, and VGG-19—are a few examples of such networks that are *off-the-shelf, downloadable,* and *pretrained networks* (Krizhevsky et al., 2012; Simonyan and Zisserman, 2014; Soekhoe et al., 2016; Szegedy et al., 2015).

Two important questions naturally arise out of using pretrained networks:

1. If we were to build a pretrained network, which dataset would we choose to learn the pretrained network's features from?
2. Do these features transfer from one dataset to another well and if so, how well?

The answer to the first question is the *neural generality* of the dataset (Venkatesan et al., 2015, 2016). The answer to the second is the *transferability of neural features* (Yosinski et al., 2014).

*Generality and Transferability*

Different datasets make a network learn different sets of filters. Consider Figure 5.2. The figure shows some handwritten character recognition datasets. Among these datasets, it is only natural for us

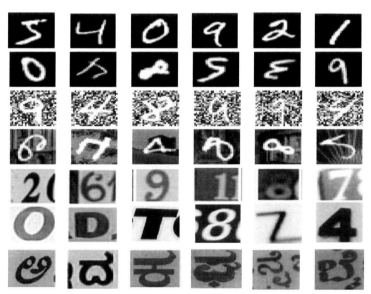

**Figure 5.2**    Handwritten character datasets: From top to bottom: MNIST (LeCun et al., 1998), MNIST-rotated (Larochelle et al., 2007), MNIST-random-background, MNIST-rotated-background, Google Street View house numbers (Netzer et al., 2011), Char74K English (de Campos et al., 2009), Char74K Kannada. Some of the images are RGB. To view the RGB image go the book's webpage at convolution.network.

to expect that MNIST-rotated contains more general features than MNIST. Due to the transformations, MNIST-rotated contains many additional structures as compared to MNIST. Hence, MNIST-rotated would require the learning of more complicated filters. For example, a network trained using MNIST-rotated will be expected to additionally have more general filters for detecting more oriented and directional edges than that using MNIST.

The filters learned from different datasets would be similar if the datasets themselves were similar. The filters we observe in the (early) layers of a network trained using a dataset represent the detectors for some common *atomic structures* in the dataset. *Atomic structures* are the forms that CNN filters take to accommodate for the entropy of the dataset it is learning on. MNIST-rotated has more entropy, therefore more atomic structures. High-entropy datasets have more varied and more complex atomic structures. In Chapter 4, we saw how MNIST leads to some simple edge detectors in the first layer. MNIST-rotated leads to more complex features and VGG has detectors for complex corners, edges, blobs, and so on, since VGG was trained on a much bigger and more complex dataset. In general, the VGG network has more atomic structures than LeNet trained on MNIST. It is fair to note that while LeNet does not have as many atomic structures as VGG, it does not need as many atomic structures as well. LeNet works with the MNIST data, which require far less atomic structures to distractively represent the data, whereas VGG works with the ImageNet data, which are notably much more complex (Figure 5.3). Even though we refer to the networks LeNet and VGG, what we imply is that these networks are trained on MNIST and ImageNet respectively. The atomic structures are emergent properties, due to the dataset the networks are trained on and not the architecture itself.

Let us perform the following thought experiment: Let us posit that all possible atomic structures constitute some space $S$. Suppose that we have a set containing the atomic structures of three datasets $D = \{D_1, D_2, D_3\}$ and $D \in S$. Assume that the dataset's atomic structures occupy manifolds as shown in Figure 5.4. It is easy to argue that $D_1$ is a more general dataset with respect to $D_2$ and $D_3$. While $D_1$ *includes* most of the atomic structures of $D_2$ and $D_3$, the latter are not as inclusive to accommodate as many atomic structures of $D_1$. Note that this thought experiment is for illustration only and does not work in a real-world setting as data manifolds are usually not clustered as described.

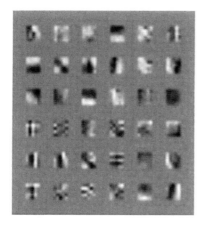 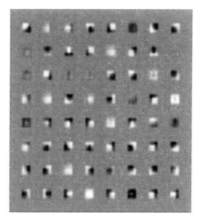

**Figure 5.3**  On the left are filters learned from Caltech 101 and on the right are VGG's first-layer filters. These filters although produced for print in Grayscale are actually in RGB. Please visit the book's webpage at convolution.network for a color rendering of the image.

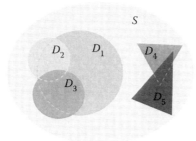

**Figure 5.4**  Thought experiment to describe the dataset generality. $S$ is the space of all possible atomic structures; $D_1$–$D_5$ are the space of atomic structures present in respective datasets.

To answer our first question of which dataset we should choose for pretraining, we need to train a pretrained network with a general enough dataset for the one we are fine-tuning with. To measure this generality in a formal manner, we need a generality metric. The generality metric should be one that allows us to compare two datasets such that we can determine which one is more general than the other, for any network architecture. A straightforward way to do that is to train a network with the base dataset, freeze the feature detector part of the network, and retrain the classifier layer alone on the new dataset. This, when compared with the performance of a randomly initialized unfrozen network on the original dataset of the same size, would give an idea as to how general the base dataset is. This was proposed in Venkatesan et al. (2015, 2016).

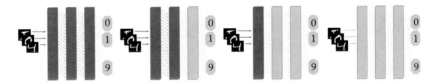

**Figure 5.5**   Freezing and unfreezing of layers during a generality metric prediction.

The basic experiment of this article is as follows: A randomly initialized network is first pretrained with a dataset. Let us call this network state $n(D_i \mid r)$, where $D_i$ represents the dataset we are initializing with and $r$ denotes the network's original configuration being random. We then proceed to retrain $n(D_i \mid r)$ in accordance with any of the setups shown in Figure 5.5.

We obtain the following network states: $n_k(D_j \mid D_i)$, which would imply that there are $k$ layers that are learned by dataset $D_j$ and were prejudiced (pretrained) by the filters of $n(D_i \mid r)$. $n_k(D_j \mid D_i)$ has $N - k$ obstinate (not allowed to learn) layers that retain and hold the prejudice of dataset $D_i$, where $N$ is the total number of layers. Obstinate layers refer to those that are frozen and ergo do not change during learning. Note that the more layers we allow to learn, imply that the network is less obstinate to learn. Also, note that these layers can be both convolutional and fully connected neural layers.

Layers learn in two facets. They learn some components that are purely their own and some that are coadapted from previous layers, which are not obstinate jointly. By making some layers obstinate, we basically fix those layers to predetermined transformations. Note that the performance gain from $n_k(D_j \mid D_i)$, and $n_{k+1}(D_j \mid D_i)$ is not just because of the new layer $k+1$ being allowed to learn, but due to the combination of all $k+1$ layers being allowed to learn.

Figure 5.5 shows the setup of these experiments. Note that in all the various obstinateness configurations, the softmax layer remains nonobstinate. The softmax layer must always be randomly reinitialized. This is because not all dataset pairs have the same number of labels or even the same label space. The unfreezing of layers happens from the rear. An unfrozen layer should not feed into a frozen layer. This is because, while the unfrozen layer learns a new filter and therefore obtains a new feature representation for the image, the latter obstinate layer is not adapting to such a transformation. When there are two layers unfrozen, the two layers should be able to coadapt together and must finally feed into an unfrozen classifier layer through unfrozen pathways only.

If the generalization performance of $n(D_j \mid r)$ is $\Psi(D_j \mid r)$ and that of $n_k(D_j \mid D_i)$ is $\Psi_k(D_j \mid D_i)$, then the dataset generality of $D_i$ with respect to $D_j$ at the layer $k$ is given by

$$g_k(D_i, D_j) = \frac{\Psi_k(D_j \mid D_i)}{\Psi(D_j \mid r)} \tag{5.1}$$

where, $g_k(D_i, D_j)$ is the generalization performance achieved by $D_j$ using $N - k$ layers worth of obstinateness from $D_i$ and $k$ layers worth of old features from $D_i$ that are combined with $k$ layers of novel knowledge from $D_j$ together. Using such a generality measure, we can now clearly make claims on which dataset is more general to pretrain a network with, while having a clear goal of our target dataset in mind. This research also showed that only some classes in a dataset alone are general enough and thus we do not have to use an entire dataset for pretraining. When we have datasets with only a few samples for some classes and a lot of samples for the others, such intraclass generality helps us train our model with only one part of the dataset first and then proceed to retrain with the whole dataset later.

To answer the second question of transferability, a study was conducted by Yosinski et al. (2014). The experiment setup was similar to that of the previous discussion but this experiment was performed using two parts of the same dataset. Instead of freezing and unfreezing the layers completely, Yosinski et al. always kept all layers unfrozen and reinitialized new layers. This experiment used one large dataset, the ImageNet, by splitting the dataset in two. The two parts of the same dataset are used to train and retrain a network like the generality experiments. They studied how transferable each layer is and how much coadaptation plays a role in learning. In their experiments, they would learn the entire network from one part of the ImageNet dataset, reinitialize part of the network again but retain the other parts, and retrain the network on the other part of the dataset. Far more interesting is the analysis of memorability. Once they retrain on the second part of the dataset, they go back and test on the first dataset again to see how much of the feature space the network remembers. To do this, they use the old classifier from the initial training on the new feature space. This gives an idea about the quality of features that cross parts of datasets. More results and interesting analysis can be found in the original paper (Yosinski et al., 2014). This article proposed a metric for transferability and showed some issues that affect negatively the transferring of features across networks trained on two parts of the same dataset. The article also studies coadaptation of layers and its role in transferability. Surprisingly, they observed that initializing a network from even a distinct task is better than initializing randomly. The article on generality was a follow-up on this work on transferability, which confirms the conclusions of the transferability paper and demonstrated further properties.

*Using Pretrained Networks for Model Compression*

In the previous section, we studied that networks that were trained on some general enough dataset could be reused as initializations for a specialized dataset and fine-tuned. In doing so, we retained the network's architecture and optimizers. The fine-tuning therefore requires computational resources similar to that of the machine in which the network was originally trained. With the reach and omnipresence of deep learning in day-to-day consumer products, it is reasonable to hope for a solution where we could expect representations capable of producing similar classification performances while needing lesser resources.

While large and deep networks can be trained reasonably efficiently on GPUs and clusters of GPUs, they have too large a memory footprint to fit in mobile phones and other consumer devices with smaller form factors. It had already been shown that in large networks, most neurons are not that useful (HasanPour et al., 2016; Li et al., 2016a,b; Nguyen et al., 2015; Wang and Liang, 2016). Most neurons are either very similar to others in their functions or do not contribute to the entropy of representation. We can therefore compress a network model so that we could make these networks portable (Bucilǔa et al., 2006). There are several other reasons why one might favor a smaller or a midsized network even though there might be a better solution available using these large pretrained networks. Large pretrained networks are computationally intensive and often have a depth in excess of 20 layers with no end in sight for deepening. A way to train larger networks faster is perhaps to learn a small network and use a technique such as Net2Net transfer learning to build networks up into bigger ones (Chen et al., 2015). Net2Net uses deterministic methods to grow networks from small to larger but the transformation is done in a *function preserving* manner so that by widening or by deepening the network, the functional mapping from one layer to another is preserved. This way once a new network is built, we can retrain it to achieve better performance. Given that the optimization and the backpropagation techniques are well defined, the worst that can happen in this situation is that the performance will not rise and remain stable. Even with these techniques, the computational requirement of these networks does not make them easily portable. Most of these networks require state-of-the-art GPUs to work even in simple feed-forward modes. This impracticality of using pretrained networks on smaller computational form factors necessitates the need to learn smaller network architectures (Srivastava et al., 2015). The quandary now is that smaller network architectures cannot produce powerful enough representations.

One method to do such compression and still retain the representations was proposed by Hinton et al. (2015). While the philosophy of using pruning and brain damage to achieve compression is a prominent research area and is showing tremendous progress, in this chapter, we will only focus on the Hintonian compression techniques (Han et al., 2015; LeCun et al., 1989; Ullrich et al. 2017). In this paper, Hinton et al. observed that the labels we backpropagate and make a network learn on is a mere fraction of what the network is actually learning. Strange as it may appear, the network seems to also be learning relationships between the classes. Hinton et al. further demonstrate that the network though trained on a specific target, has the tendency to learn knowledge that is not supervised by this target. Using the idea that networks learn interclass relationships in their representations, they were able to create richer targets (with these interclass relations) for a much smaller network to learn on. Since the supervision has much more entropy, the networks need a smaller number of neurons to generalize. We shall study model compression using this idea in this section.

Even though we have assumed that the classes are independent from each other and each class produces its own loss, this is not strictly true. In MNIST, for instance, some twos look like sevens and some nines look like sixes and so on. Although not explicitly supervised, the network itself can form these relationships and comparisons between the classes. Consider for instance the case shown in Figure 5.6. The original hard targets are the labels. In this case, we are learning a *Dog*, so the label bit representing *dog* is 1 and for the others is 0. This is what we

| Cow | Dog | Cat | ... | Type of labels |
|---|---|---|---|---|
| 0 | 1 | 0 | ... | Hard labels |
| 0.000006 | .9 | 0.99991 | ... | Ensemble |
| 0.05 | 0.3 | 0.2 | ... | Soft labels |

**Figure 5.6**   Softening the output probabilities.

would expect out of a softmax, a probability close to 1 for dog and close to 0 for the rest. Assume that there is a large network or an ensemble of networks that already learned on this dataset and once we feed-forward the *dog* image through it, it produces the output shown in the geometric ensemble of Figure 5.6. In this ensemble softmax outputs, what is obviously noticeable is that the probability for dog is very high, the probability for cat is the next highest although it is only 10% of a dog, and the probability of other labels are subsequently lower, although perhaps well within the precision of the floating point. What is noticeable is that the second maximal probability is not such a bad option after all. Among the other choices, cat looks the closest to a dog. Similar effects were also seen in the AlexNet outputs from Chapter 4.

Hinton et al. formed a new perspective of these softmaxes in that the network learns implicitly that some classes are like others, even though such class–class similarities were never provided as supervision. In this instance, even though we have not created any special labeling for the fact, the network learns that there is a small chance that a cat may look like a dog. From the perspectives of distributions of data spread on the feature space as discussed in Chapter 1, we can imagine this as follows. A network learning discriminative representations of classes does in a manner such that classes that are related to each other, are closer to each other (even overlapping slightly). This information that we did not specifically label but the network learns by itself is referred to as *dark knowledge*.

In this paper, a *temperature* softmax was used instead of the regular softmax. A temperature softmax is defined for class $i$ as

$$p_i = \frac{\exp\left(\dfrac{z_i}{T}\right)}{\sum_j \exp\left(\dfrac{z_j}{T}\right)} \tag{5.2}$$

where $T$ is the temperature. This transformation smoothes out the ensemble predictions and makes the differences more prominent. $T = 1$ will be the normal softmax and a higher temperature would spread the probabilities out. The higher the temperature, the lower the difference between the two classes in terms of their softmax. By using this, we can see clearly what the network is guessing as its second and later options. Figure 5.6 shows this softening of labels through a temperature-raised softmax.

This temperature softmax output provides insights as to what kind of connections exists between classes. Consider now the idea of fine-tuning from the previous section. We want to do this more efficiently or with lesser number of neurons or layers while still having the ability to produce the same representations. To achieve this, we need additional supervision than what is available from our dataset itself. We can use the temperature-raised softmax outputs of the first pretrained network as additional supervision to guide the smaller network along with the dataset's own loss.

For typical fine-tuning (to be precise training as this will be a new randomly initialized network) on a smaller network, we use backpropagation of the errors on the hard labels. In this case, we can provide the network additional supervision if we backpropagate the error to these temperature-raised soft labels from another network along with the usual backpropagation of the hard labels. Training with soft targets alone is not enough to produce a strong gradient, so in practicality we would create an error function that is a weighted combination of the soft and hard targets. One may think of this as dark knowledge transfer. Suppose $p_T(i)$ is the temperature softmax outputs of the $i$th batch of inputs to the parent network $p$, $y(i)$ is the true label, and $c(i)$ is the softmax and $c_T(i)$ the temperature-raised softmax of the child network that is currently learning. Let the loss be $\|c(i) - y(i)\|$ and the difference between the networks' temperature softmax outputs $\|c_T(i) - p_T(i)\|$ is for the batch. The new loss that is backpropagated is

$$e(w) = \|c(i) - y(i)\| + \alpha \|c_T(i) - p_T(i)\| \qquad (5.3)$$

In their research with only a loss of 0.02%, the authors could transfer dark knowledge from a 1200-1200-10 network to an 800-800-10 network for the MNIST dataset. More surprising is the result of a follow-up experiment where they withheld the entire sample space of a particular class and were still able to learn it by merely using the dark knowledge transfer. For instance, if all the samples with original labels 3 were withheld from training the child network and only soft targets were propagated for the class of 3, the network was still able to perform well for the class of 3.

This technique is also often referred to as model compression or, as noted above, it is colloquially referred to as dark knowledge transfer. Dark knowledge transfer is also a type of a strong regularizer. Unlike $L_1$ and $L_2$ regularizers that directly penalize the weights for being too large or too nonsparse, dark knowledge from another network penalizes the network for not producing outputs that are similar in pattern to the previous network. This is also looked at as modeling the distribution of outputs of the parent network by the child network. Using this technique, it was shown that smaller networks could potentially be trained that can map an image to the same label space just as effectively as a larger network.

Nowadays, many computer vision architects prefer to not train a new architecture from scratch and instead prefer to fine-tune a pretrained network whenever computational resources are a constraint. However, training even a midsized network with a small dataset is a notoriously difficult task. Training a deep network, even those with midlevel depth, requires a lot of supervision in order to avoid weight explosion. In these cases, dark knowledge transfer helps in providing some additional supervision and regularization.

*Mentee Networks and FitNets*

Taking the dark knowledge idea further, Romero et al. also tried to backpropagate some intermediate layers' activations along with the soft labels (Romero et al., 2014). This provides an even more appealing prospect of adding additional supervision. The networks that receive such supervision are often referred to as *mentee* networks and the networks that provide the supervision, the mentor networks. The authors in this paper originally used this idea to produce a thinner and deeper version of the parent network. A deeper network is more productive than a wider network when it comes to CNNs and other neural networks. Evidence for this claim has been building since before AlexNet (LeCun et al., 2015). If we were to have two networks of different configurations, we learn the same label mapping using dark knowledge transfer; the next step is to transfer not just the softmax layers but also the intermediate layers.

Mentee nets were first proposed as *fitNets* to learn thinner but deeper networks by Romero et al. (2014). Several works generalized this idea (Chan et al., 2015; Gupta et al., 2015; Venkatesan et al., 2015, 2016). A recent paper also used this as a good initializer (Tang et al., 2015). Another used this technique to transfer knowledge between two different types of networks, in this case a recurrent neural network (RNN) and a CNN (Geras et al., 2015; Shin et al., 2016). The principle idea of mentor nets is to propagate activations and make another network learn a feature space of the same or a similar capability. Therefore, smaller networks now will be capable of representing the feature spaces and intermediate representations of a larger network.

Figure 5.7 describes the connections required to learn such a system. The top network is a pretrained off-the-shelf network and the bottom one is a randomly created and initialized network of a different configuration. They both feed-forward the same data and it is expected that they produce the same softmax output (using dark knowledge transfer). In doing so, we also want to produce similar activations across some layers deep in the network. Note that this connection is required only during the training process. Once trained, the smaller mentee network is capable of working on its own accord.

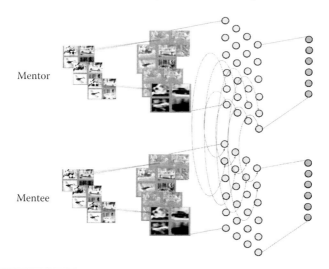

**Figure 5.7**   Mentor–mentee connections.

Also, no backpropagation is done on the larger mentor network, but only a forward propagation. Backpropagation is performed only on the smaller mentee networks.

A typical weight update for this setup would look like the following:

$$w^{t+1} = w^t - \eta \left[ \alpha_t \frac{\partial}{\partial w} e(w) + \beta_t \sum_{\forall (l,j) \in B} \frac{\partial}{\partial w} \psi(l, j) + \gamma_t \frac{\partial}{\partial w} \psi(n, m) \right]$$

$$(5.4)$$

where $n$ and $m$ are the number of layers (including softmax layers in the network) and $B$ includes the pairs of layers from mentor to mentee to be supervised. The first term inside the square brackets is the typical update of the error in a normal stochastic gradient descent, the second term is the weight update for the mentorship, and the last term is the weight update for the dark knowledge transfer. Various configurations of $\alpha$, $\beta$, and $\gamma$ control how learning happens. It was shown that using various learning rates, several different characteristics of these mentee networks shall be produced and one can even learn the exact feature space of the mentor network directly (Romero et al., 2014; Venkatesan et al., 2015, 2016).

In the experiments performed with the Caltech101 and 256 datasets, it was found that the mentee networks performed better than the vanilla network. The mentee network was also able to perform significantly better than the independent network when only the classifier/multilayer perceptron (MLP) sections were allowed to learn on the Caltech-256 dataset with representation learned from Caltech101. This proves the generality of the feature space learned.

*Application Using Pretrained Networks: Image Aesthetics Using CNNs*

Let us posit a network pretrained on some image categorization task and study its application on a different task. The task we are considering now is that of comparing the aesthetics of images, which is drastically different from that of categorization (Chandakkar et al., 2017).

CNNs have achieved success in tasks where objective evaluation is possible. However, there are tasks that are inherently subjective, for example, estimation of image aesthetics. CNNs have also shown impressive results in such tasks, outperforming approaches based on handcrafted features that were inspired from facts derived from the photography literature. It is possible that CNNs have discovered some yet undefined attributes essential to the appearance of the images (Lu, 2014, 2015).

Automated assessment of image aesthetics using pretrained AlexNet CNNs was first proposed in Lu (2014). RAPID contains a two-channel CNN, which essentially is two AlexNets bundled into one. One CNN takes the entire image as input whereas the other one takes a randomly cropped region as input. Finally, feature vectors of these two inputs (the representation of the last layer) are combined and then a new regressor is initialized at the end to produce an aesthetics score for an image.

Given the subjectivity of this task, one could argue that assigning an absolute score to each image is difficult and hence is prone to errors. On the other hand, comparing aesthetic qualities of two images may be more reliable. This too has practical applications. Often it happens that a user has a collection of photos and the user is only interested in ranking them instead of getting the scores for individual photos. In other applications, such as image retrieval based on aesthetics and image enhancement, this ranking-based approach is necessary. This was the motivation behind the approach proposed in Chandakkar et al. (2017).

The architecture behind the key idea of the work in Chandakkar et al. (2017) is shown in Figure 5.8. It takes a pair of images as input. Each image is processed using an architecture similar to the RAPID method. The feature vectors of both images are called $C_1$ and $C_2$, respectively, and are also shown in Figure 5.8. $C_1$ and $C_2$ are processed accordingly to produce a score denoting the aesthetics ranking between two images. The loss function takes the form of $\max(0, )$. It will produce a loss only if the ranking of two images is different from the ground truth. Otherwise, the loss produced is zero. Thus, without much training on the representation learning system of AlexNet, we could apply it to a completely different task of aesthetics.

Applications like these demonstrate that even though the AlexNet is trained on some image categorization task, the learning process might have allowed the network to figure out features that are general enough so that they could be used in other types of tasks that may appear to be completely different. We mention in passing that another good example

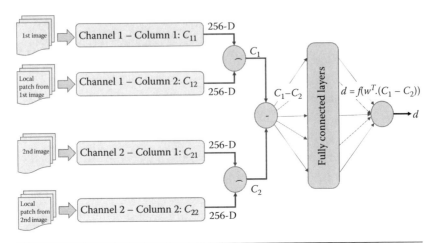

**Figure 5.8** Architecture of the aesthetics network; weights are shared between columns $C_{11}$ and $C_{21}$, $C_{12}$ and $C_{22}$; the features obtained from $C_{11}$ and $C_{12}$ are concatenated (represented by $\frown$ symbol) to get $C_1$, and $C_{21}$ and $C_{22}$ are concatenated to get $C_2$; the vector $C_1-C_2$ is passed through two dense layers to obtain a score $d$ comparing the aesthetics of two images. $f()$ denotes a ReLU nonlinearity. Please refer to the text for further details. (Courtesy of Parag Chandakkar.)

of this sort is the *video2vec* framework proposed in Hu et al. (2016), where the pretrained CNNs were used in conjunction with RNNs to learn a semantic spatial-temporal embedding space for videos so that typical video analysis tasks like action recognition can be better supported.

### Generative Networks

The networks that we studied thus far all try to minimize a loss in predicting a distribution over categorization labels. These are commonly referred to as discriminator networks. These networks sample data from a distribution in a high-dimensional space such as images and map it to a label space or some space of feature representations. There is another class of networks that are a corollary to these networks called generative networks. Generative networks sample data from a distribution on a latent space, which is typically much smaller than the space of the data, and generate samples that mimic samples drawn from the distributions of a high-dimensional space. In the interest of computer vision, these networks sample from a distribution of latent space and generate images back. In this section, we study two of the most popular models of generative networks.

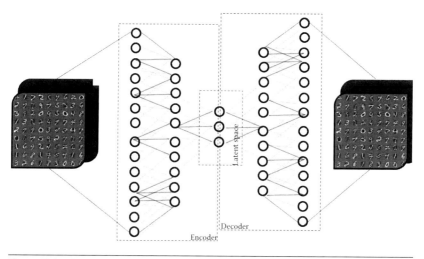

**Figure 5.9**   Autoencoder.

*Autoencoders*

Autoencoders are perhaps the most prominent case study for CNN networks not being used for categorization. Autoencoders as shown in Figure 5.9 are mirrored symmetric networks that learn to regenerate the image themselves. The autoencoder consists of two parts: the encoder and the decoder. The encoder creates a succinct and compact representation of the image, often referred to as the codeword or the latent space representation. The decoder reconstructs the image back from these latent space representations. Autoencoders are comparable to dictionary learning or to principal component analysis (PCAs) in the sense that they create a representation that is closer to the true underlying structure of the data independent of the categorization labels and in a lower (most common) dimensional space. Typically, the decoder is not decoupled from the encoder. Often the decoder weights are just a transpose of the encoder weights, therefore decoders and encoders are symmetric and share weights. If the layers were convolutional on the encoder, we may use a transposed convolution or fractionally strided convolutions. These are also (with some abuse of terminology) referred to as deconvolutions. These are quite simply the gradients of some convolution operation, which is a convolution with the transposed filters. For maxpooling, we can use strided convolutions to unpool or store the locations of the maxpool to unpool. These are referred to as fractionally strided convolutions and they upsample the signal before convolving. One may think of this as akin to learning a custom upsampling operation that mimics the unpooling operation (Zeiler et al., 2010).

An auto encoder has two parts, the encoder $E(x) = c$, which takes and input $x$ and produces a codeword representation of the image $c$ and the decoder $D(c) = \hat{x}$, which takes in as input the codeword $c$ and produces a reconstruction of the image $\hat{x}$. Since the weights are usually *tied* between the encoder and the decoder, we need to learn the weights of either $E$ or $D$. We do it by minimizing the reconstruction error between $\hat{x}$ and $x$. We may choose to use any of the previous error to accomplish this, but RMSE errors are typically preferred. Figure 5.10 shows the typical input-codeword-reconstructions of a simple one-layer autoencoder. The code for this autoencoder from which the figures were created is also available at the Yann tutorials (refer to Appendix A).

Autoencoders come in many forms and traits. Autoencoders that use neural networks are either overcomplete or undercomplete. If the dimensionality of the codeword is less than that of the input, it is an undercomplete autoencoder; if otherwise, it is overcomplete. Overcomplete autoencoders often need strong regularization and sparsity imposed on them to be able to learn. Since the image is reconstructed from an embedded subspace, the reconstruction is not expected to be perfect. While this may appear to be a drawback initially, the autoencoder could be used to generate images where we deliberately require a nonperfect reconstruction. Learning overcomplete autoencoders is very difficult. Adding noise to the input and expecting to reconstruct a noise-free image is a strong way of regularizing these networks so that we can easily learn larger and deeper networks.

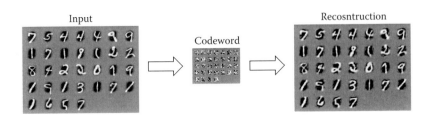

**Figure 5.10**   Encoding and decoding MNIST using a one hidden-layer autoencoder.

Consider for instance the problem of denoising. In denoising, we want to remove the noise from an image. While autoencoders learn to reconstruct an image using some structures from the images, these structures are often not found in noisy images. Encoders encode only the statistically common features among the images. Noise is a stochastic process and therefore is not a common structure among the datasets. The latent space for a noisy image and one without noise will be close to being the same (if not the same). Therefore, regenerating a noisy image that is feed-forward through an autoencoder will lead to the generation of potentially clean images. Denoising autoencoders is among the most popular implementations of autoencoders (Vincent et al., 2008, 2010). It turns out that learning an autoencoder with a corrupted set of inputs is better than learning with a clean set of inputs as well. This is because learning features form a corrupted set of features that help the encoder learning better and more robust representations than simply learning an identity. Corrupting the input to the autoencoder is very similar to the dropout regularizer that we previously studied. In fact, one of the noises that was added to the input was making random bits zero. One might as well claim that Vincent et al. proposed the first dropout algorithm.

The encoders of autoencoders are also used as initialization for discriminative networks. Autoencoders are generative methods, therefore they learn fundamental features pure and untouched by the entropy of class labels. Therefore, the features learned from them deal only with the structure in the image. Networks learned thus, as either mentoring or as fully pretrained network initializations, help in establishing a strong initialization for discriminative networks. Autoencoders in their various manifestations are referred to as unsupervised learning and are an active area of study and research (Bengio, 2012; Bengio et al., 2012; Erhan et al., 2010; Goodfellow et al., 2014; Radford et al., 2015).

*Generative Adversarial Networks*

Autoencoders generate samples that tend to be close to the mean of the samples that are represented by the latent representations. To be able to better sample at high dimensionality, Goodfellow et al. devised a new class of generative networks called adversarial networks (Goodfellow et al., 2014). Unlike the previous networks, these networks do not have an external objective, but the objective itself is internal to the machine. Adversarial generative learning involves two networks. One of the networks is a generator that is a typical decoder of the type discussed in the previous section. The generator samples from a random latent distribution and generates an image. The second network is a discriminator that tries to predict if an input image was generated by the generator network or was sampled from a dataset. In case the data were sampled from the real world, we may also train a softmax layer with the final layer of the discriminator's representations to learn the features in a discriminative way.

The GAN works in the following manner. Consider that the data is distributed along $p_{data}$, whose form we do not know. Consider though that we assume that $p_{data}$ is a parameterized distribution of some form $p_{model}$. In a maximum likelihood setting, we would want to find a good set of parameters so that we cannot differentiate between samples generated from $p_{model}$ and $p_{data}$. We can do this by doing the following maximum likelihood estimate of the parameters $w$.

$$\ddot{w} = \text{argmax}_w \; \mathbb{E}_{x \to p_{data}} \log p_{model}(x \mid w). \tag{5.5}$$

In GANs, we have two networks, the generator $G(z)$ that takes as input a random vector $z$ and transforms it into an image, and the discriminator $D(x)$, which takes an input image $x$ and produces a probability that $x$ is real as against produced by $G$. The GAN system is essence a game played by $G$ and $D$. $D$ adjusts its weights with the goal of trying to differentiate between $x$ from the real-world data and $G(z)$. $G$ adjusts its weight so as to produce $G(z)$ so realistic that $D$ claims $D(G(z)) = 1$ (the generated sample is real). We abuse the notation a little here by referring to the probability produced by $D$ as $D(x)$ or $D(G(z))$. The GAN game is the following two steps. For $D$,

$$\min_D - \mathbb{E}_{x \to p_{data}(x)}\Big[\log\big(D(x)\big)\Big] - \mathbb{E}_{z \to p_z(z)}\Big[\log\big(1 - D(G(z))\big)\Big] \tag{5.6}$$

where $D$ tries to minimize $D(G(z))$ and maximize $D(x)$, implying that it is learning to differentiate between fake and real images. For $G$,

$$\min_G - \mathbb{E}_{z \to p_z(z)}\Big[\log\big(D(G(z))\big)\Big] \tag{5.7}$$

where $G$ learns to produce $G(z)$ that would have made $D$ predict a high value for $D(G(z))$. It is to be noted that while optimizing

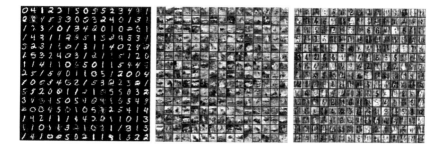

**Figure 5.11**    Some sample images generated from MNIST, CIFAR10 and SVHN images using classes 0–5 of the three datasets using very simple LeNet-sized **G** and **D**. These images are produced in grayscale and color versions of these images are available at the book's webpage at convolution. network. Furthermore, the code that generated these images are also available at yann tutorials.

Equation (5.6), only the weights of $D$ are updated and while optimizing Equation (5.7), only the weights of $G$ are updated. This implies that $G$ never observes any image and still learns to generate images. Figure 5.11 shows some of the generated images.

These networks are trained simultaneously in this minimax game sort of a way and there is some theoretical evidence to suggest that an equilibrium state could be reached where the discriminator will no longer be able to predict whether the data originated from the real world or from the generator (Goodfellow et al., 2014). This converges the learning. At convergence, {D} always produces the value 0.5 irrespective of where the input is sampled from. Figure 5.12 shows the architecture for a typical generative adversarial network (GAN).

The training of GANs typically begins with the discriminator. The discriminator may even first be preliminarily trained to recognize the samples from the dataset using the softmax layer. Once the discriminator is in place, we start feeding it the samples generated by the thus far untrained generator. The discriminator produces an error at the classifier for predicting if the image came from the dataset or from the generator. As the learning proceeds, the generator network learns to produce samples that are closer and closer to the original data up to a point where the generation is indistinguishable (at least for the discriminator network) from the data. For producing better images, we may use the transposed convolutions and the extension made using deconvolutional GANs (Radford et al., 2015).

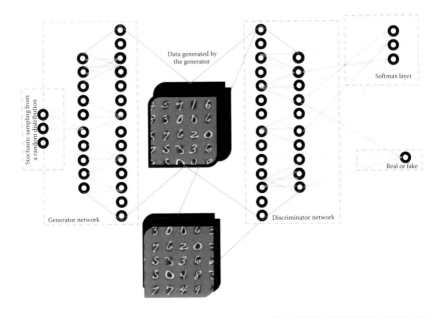

**Figure 5.12**   Generative adversarial networks.

A problem with GANs is that the generator network has the tendency to get stuck in the same mode. This implies that the generator does not capture the diversity of the entire dataset but locks on to only one type of data. In Figure 5.11 we can already notice that some classes are produced more often than others. For instance, consider the situation in Figure 5.13. The circles with the dots are the original dataset's distribution. The task of the GAN is to produce data that appear to be sampled from this original dataset. A well-trained GAN would appear to sample data from the distribution that is represented by the square-filled area. This GAN would be able to produce samples of all modes or types. A GAN that is stuck in one mode would produce samples from a distribution that is unimodal (or a few modes). This will lead to producing the same or at least similar data again and again. One way to solve this problem is by using mode-regularized GAN (MRGAN) (Che et al., 2016). In the MRGAN, the authors introduce new ways of regularizing the objective that stabilize the training of GAN models. They demonstrate that such regularizers help distribute the probability mass across many modes of the distributions that generate the data.

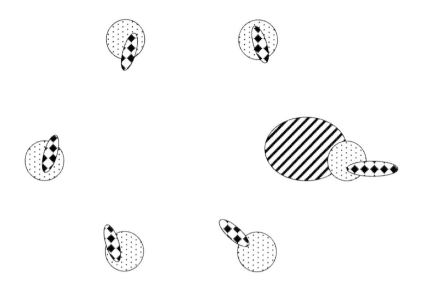

**Figure 5.13**   A GAN gets stuck in one mode. The five circles that are filled with dots are the original distribution of the dataset. The ellipses with the same filling represent the distribution of one generator. The generator with the slanted line filling is stuck in one mode and the one filled with squares is a well learned, albeit not perfect generator.

One recent addition to GANs is to make these generator representations interpretable. This was accomplished by using disentangled representations and InfoGAN (Chen et al., 2016). InfoGAN tried to maximize the mutual information between a small subset of the latent variables and the generation. The generator samples from not only a random distribution but also from a controlled latent space that describes some properties about the sample that is being learned. This architecture is described in Figure 5.14. This allows the network gradients to maximize the information provided by these variables with generation. In the simplest possible form, the generation is the reverse of classification and that it generates samples of a class that was requested implies that the generator is trained discriminatively. Several proof-of-concept experiments were performed to demonstrate this effect that could be found in the article (Chen et al., 2016). There are several iterations of GANs that have been discovered since its original formulation. Least squares GAN and Wasserstein GANs are some popular modifications (Arjovsky et al., 2017; Mao et al., 2016).

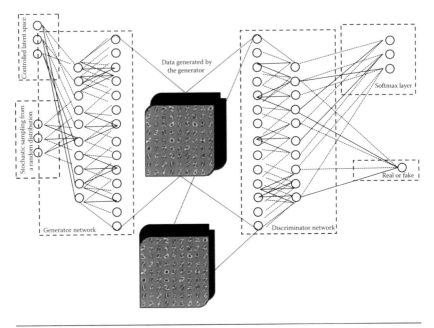

**Figure 5.14**   InfoGAN architecture.

A particularly smart thing about GANs is that they do not directly encode or compress data in the literal sense, but rather sample from a model of a distribution that it has learned. This implies that the generator of a GAN is a source of a datastream. This is particularly useful for many applications including incremental learning (Venkatesan et al. 2017). GANs are a new leap in neural networks in what is seemingly a treasure trove of potential.

## Summary

In this chapter, we looked at several modern and novel usages of CNNs, each giving us a novel perspective on how deep networks could be used. Deep learning is being used on a multitude of problems in more creative ways every day. Products developed using deep learning technology span a large spectrum. Starting from consumer products such as image searching and voice recognition to more somber medical applications, deep learning has become a ubiquitous and integral part of modern technology. While there exists a wide swath of applications, in this chapter we chose to study briefly those that lend the reader a new perspective on the use and application of these networks.

We noticed that most of these could be accomplished by using the loss functions and activations differently. We discussed how datasets of images might hold some similar and general properties that might be useful in deciding which dataset to use when training. We studied the transferability of deep learning where we noticed that some networks learned on one dataset could be used to learn from another dataset. We observed that the networks learn much more than what they were explicitly made to learn. We used this to our advantage and tried to learn a class distribution with very samples of that class. We extended this beyond simple dark knowledge transfer and made one network *teach* another.

We also discussed generative learning where we recreate the image back from some middle layer, which we called a *codeword* layer. We noticed that this imperfect recreation is useful in cases such as denoising. CNNs are full of potential and newer and novel usages of these networks are being discovered every day and applied to a variety of problems helping humanity.

## References

Arjovsky, Martin, Chintala, Soumith, and Bottou, Léon. 2017. Wasserstein gan. arXiv preprint arXiv:1701.07875.

Bengio, Yoshua. 2012. Deep learning of representations for unsupervised and transfer learning. *ICML Unsupervised and Transfer Learning* 27: 17–36.

Bengio, Yoshua, Courville, Aaron C, and Vincent, Pascal. 2012. *Unsupervised feature learning and deep learning: A review and new perspectives*. CoRR, abs/1206.5538, 1.

Buciluă, Cristian, Caruana, Rich, and Niculescu-Mizil, Alexandru. 2006. Model compression. *Proceedings of the 12th ACM SIGKDD International Conference on Knowledge Discovery and Data Mining*, pp. 535–541, San Fransisco, California. ACM.

Chan, William, Ke, Nan Rosemary, and Lane, Ian. 2015. *Transferring knowledge from a RNN to a DNN*. arXiv preprint arXiv:1504.01483.

Chandakkar, Parag, Gattupalli, Vijetha, and Li, Baoxin. 2017. A computational approach to relative aesthetics. arXiv preprint arXiv:1704.01248.

Che, Tong, Li, Yanran, Paul Jacob, Athul et al. 2016. *Mode regularized generative adversarial networks*. arXiv preprint arXiv:1612.02136.

Chen, Tianqi, Goodfellow, Ian, and Shlens, Jonathon. 2015. Net2Net: Accelerating learning via knowledge transfer. arXiv preprint arXiv:1511.05641.

Chen, Xi, Duan, Yan, Houthooft, Rein et al. 2016. InfoGAN: Interpretable representation learning by information maximizing generative adversarial nets. *Advances in Neural Information Processing Systems*, pp. 2172–2180, Barcelona, Spain. NIPS.

de Campos, TE, Babu, BR, and Varma, M. 2009. Character recognition in natural images. *Proceedings of the International Conference on Computer Vision Theory and Applications,* Lisbon, Portugal.

Erhan, Dumitru, Bengio, Yoshua, Courville, Aaron et al. 2010. Why does unsupervised pre-training help deep learning? *Journal of Machine Learning Research* 11: 625–660.

Everingham, Mark, Van Gool, Luc, Williams, Christopher KI et al. n.d. *The PASCAL Visual Object Classes Challenge 2012 (VOC2012) results.* Retrieved from PASCAL VOC: http://www.pascal-network.org/challenges/VOC/voc2012/workshop/index.html

Fei-Fei, Li, Fergus, Rob, and Perona, Pietro. 2006. One-shot learning of object categories. *IEEE Trans. Pattern Recognition and Machine Intelligence* (IEEE) 28: 594–611.

Geras, Krzysztof J, Mohamed, Abdel-rahman, Caruana, Rich et al. 2015. *Compressing LSTMs into CNNs.* arXiv preprint arXiv:1511.06433.

Goodfellow, Ian, Pouget-Abadie, Jean, Mirza, Mehdi et al. 2014. Generative adversarial nets. *Advances in Neural Information Processing Systems,* pp. 2672–2680.

Gupta, Saurabh, Hoffman, Judy, and Malik, Jitendra. 2015. *Cross modal distillation for supervision transfer.* arXiv preprint arXiv:1507.00448.

Han, Song, Mao, Huizi and Dally, William. 2015. Deep compression: Compressing deep neural networks with pruning, trained quantization and huffman coding. arXiv preprint arXiv:1510.00149.

HasanPour, Seyyed Hossein, Rouhani, Mohammad, Vahidi, Javad et al. 2016. *Lets keep it simple: Using simple architectures to outperform deeper architectures.* arXiv preprint arXiv:1608.06037.

Hinton, Geoffrey, Vinyals, Oriol, and Dean, Jeff. 2015. *Distilling the knowledge in a neural network.* arXiv preprint arXiv:1503.02531.

Horn, Berthold KP and Schunck, Brian G. 1981. Determining optical flow. *Artificial Intelligence* (Elsevier) 17: 185–203.

Hu, Sheng-heng, Li, Yikang, and Li, Baoxin. 2016. Video2Vec: Learning semantic spatial-temporal embeddings for video representation. In *International Conference on Pattern Recognition,* Cancun, Mexico.

Karpathy, Andrej and Fei-Fei, Li. 2015. Deep visual-semantic alignments for generating image descriptions. *Proceedings of the IEEE Conference on Computer Vision and Pattern Recognition,* pp. 3128–3137, Santiago, Chile. IEEE.

Krizhevsky, Alex, Sutskever, Ilya, and Geoffrey, Hinton E. 2012. ImageNet classification with deep convolutional neural networks. *Advances in Neural Information Processing Systems,* Harrahs and Harveys, Lake Tahoe: NIPS, pp. 1097–1105.

Larochelle, Hugo, Erhan, Dumitru, Courville, Aaron et al. 2007. An empirical evaluation of deep architectures on problems with many factors of variation. *Proceedings of the 24th International Conference on Machine Learning,* pp. 473–480, Corvallis, Oregon. ACM.

LeCun, Yann, Bengio, Yoshua, and Hinton, Geoffrey. 2015. Deep learning. *Nature* 52: 436–444.

LeCun, Yann, Bottou, Leon, Bengio, Yoshua et al. 1998. Gradient-based learning applied to document recognition. *Proceedings of the IEEE* 86: 2278–2324.

LeCun, Yann, Denker, John S, Solla, Sara A, Howard, Richard E and Jackel, Lawrence D. 1989. Optimal brain damage. *NIPs* 2: 598–605.

Li, Hao, Kadav, Asim, Durdanovic, Igor et al. 2016a. *Pruning filters for efficient ConvNets*. arXiv preprint arXiv:1608.08710.

Li, Yikang, Hu, Sheng-hung, and Li, Baoxin. 2016b. Recognizing unseen actions in a domain-adapted embedding space. *2016 IEEE International Conference on Image Processing (ICIP)*, pp. 4195–4199, Phoenix, Arizona. IEEE.

Lu, Xin, Lin, Zhe, Jin, Hailin et al. RAPID: Rating pictorial aesthetics using deep learning. *Proceedings of the 22nd ACM International Conference on Multimedia*, pp. 457–466. Orlando, FL, ACM.

Lu, Xin, Lin, Zhe, Shen, Xiaohui et al. 2015. Deep multi-patch aggregation network for image style, aesthetics, and quality estimation. In *Proceedings of the IEEE International Conference on Computer Vision*, pp. 990–998.

Mao, Xudong, Li, Qing, Xie, Haoran, Lau, Raymond YK, Wang, Zhen, Smolley, Stephen Paul. 2016. Least squares generative adversarial networks. arXiv preprint ArXiv:1611.04076.

Netzer, Yuval, Wang, Tao, Coates, Adam et al. 2011. Reading digits in natural images with unsupervised feature learning. *NIPS Workshop on Deep Learning and Unsupevised Feaure Learning.*

Nguyen, Anh, Yosinski, Jason, and Clune, Jeff. 2015. Deep neural networks are easily fooled: High confidence predictions for unrecognizable images. *Proceedings of the IEEE Conference on Computer Vision and Pattern Recognition*, pp. 427–436, Boston, MA. IEEE.

Radford, Alec, Metz, Luke, and Chintala, Soumith. 2015. *Unsupervised representation learning with deep convolutional generative adversarial networks.* arXiv preprint arXiv:1511.06434.

Romero, Adriana, Ballas, Nicolas, Kahou, Samira et al. 2014. *FitNets: Hints for thin deep nets.* arXiv preprint arXiv:1412.6550.

Russakovsky, Olga, Deng, Jia, Su, Hao et al. 2015. Imagenet large scale visual recognition challenge. *International Journal of Computer Vision* 115: 211–252.

Shin, Sungho, Hwang, Kyuyeon, and Sung, Wonyong. 2016. *Generative transfer learning between recurrent neural networks.* arXiv preprint arXiv:1608.04077.

Simonyan, Karen and Zisserman, Andrew. 2014. *Very deep convolutional networks for large-scale image recognition.* arXiv preprint arXiv:1409.1556.

Soekhoe, Deepak, van der Putten, Peter, and Plaat, Aske. 2016. On the impact of data set size in transfer learning using deep neural networks. *International Symposium on Intelligent Data Analysis,* pp. 50–60.

Srivastava, Rupesh K., Greff, Klaus, and Schmidhuber, Jurgen. 2015. Training very deep networks. *Advances in Neural Information Processing Systems*, pp. 2377–2385. NIPS.

Szegedy, Christian, Liu, Wei, Jia, Yangqing et al. 2015. Going deeper with convolutions. *Proceedings of the IEEE Conference on Computer Vision and Pattern Recognition*, pp. 1–9, Boston, MA. IEEE.

Tang, Zhiyuan, Wang, Dong, Pan, Yiqiao et al. 2015. *Knowledge transfer pre-training*. arXiv preprint arXiv:1506.02256.

Ullrich, Karen, Meeds, Edward and Welling, Max. 2017. Soft weight-sharing for neural network compression. arXiv preprint arXiv:1702.04008.

Venkatesan, Ragav, Chandakkar, Parag, and Li, Baoxin. 2015. Simpler non-parametric methods provide as good or better results to multiple-instance learning. *Proceedings of the IEEE International Conference on Computer Vision*, Sanitago, Chile: IEEE, pp. 2605–2613.

Venkatesan, Ragav, Gattupalli, Vijetha, and Li, Baoxin. 2016. On the generality of neural image features. *IEEE International Conference on Image Processing*. Phoenix: IEEE.

Venkatesan, Ragav, Venkateswara, Hemanth, Panchanathan, Sethuraman, and Li, Baoxin. 2017. A strategy for an uncompromising incremental learner. arXiv preprint arXiv:1705.00744.

Vincent, Pascal, Larochelle, Hugo, Bengio, Yoshua et al. 2008. Extracting and composing robust features with denoising autoencoders. *Proceedings of the 25th International Conference on Machine Learning*, pp. 1096–1103, Helsinki, Finland. ACM.

Vincent, Pascal, Larochelle, Hugo, Lajoie, Isabelle et al. 2010. Stacked denoising autoencoders: Learning useful representations in a deep network with a local denoising criterion. *Journal of Machine Learning Research* 11: 3371–3408.

Wang, Xing and Liang, Jie. 2016. *Scalable compression of deep neural networks*. arXiv preprint arXiv:1608.07365.

Wasserman, Philip D. 1993. *Advanced Methods in Neural Computing*. New York, NY: John Wiley & Sons.

Yosinski, Jason, Clune, Jeff, Bengio, Yoshua et al. 2014. How transferable are features in deep neural networks? *Advances in Neural Information Processing Systems*, pp. 3320–3328.

Zeiler, Matthew D, Krishnan, Dilip, Taylor, Graham W et al. 2010. Deconvolutional networks. *Computer Vision and Pattern Recognition (CVPR), 2010 IEEE Conference*, pp. 2528–2535, San Fransisco, CA IEEE.

# A
## Yann

Along with this book, we are delighted to note that we are launching a toolbox to build and learn convolutional neural networks and other networks that were studied in this book. The toolbox is called the *yet another neural network* (yann) *toolbox* and is hosted at http://www .yann. network. The yann toolbox is built on Python 2.7 and on top of the Theano framework with future plans to be able to use both the Theano and TensorFlow frameworks (Bergstra et al., 2010; Abadi et al., 2016). Yann is very simple to install and setup. All one needs to do is run the following command on Linux for a simple version of the installer:

```
>> pip install git+git://github.com/ragavvenkatesan/
   yann.git
```

Yann is in its early phases and is presently undergoing massive development. Although some changes are expected to be made, this chapter presents only the core of the toolbox, which should remain valid even if the actual toolbox may keep evolving. If there are indeed changes, the release notes of each version and its API documentation would reflect it. This appendix corresponds to version 1.0rc1, which is the release candidate for Yann version 1. There will always be an up-to-date tutorial version of this chapter available with the toolbox documentation. While there are more formal and wholesome toolboxes that have a much larger userbase such as ConvNet, Lasagne, Keras, Blocks, and Caffe (Lasagne, n.d.; Chollet, 2015; Vedaldi and Lenc, 2015; Jia et al., 2014), this toolbox is much simpler and versatile for a beginner to learn quickly. Hence, yann is the perfect supplement to this book. It is also a good choice for a toolbox for running pretrained models and builds complicated, nonvanilla architecture that is not easy to build with the other toolboxes. Yann was also used in the deep learning for computer vision course (available at course.convolution.network) that was developed from this book and taught at ASU in the Spring of 2017.

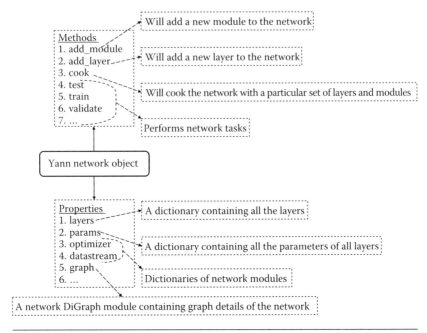

**Figure A.1** The structure of the network object.

## Structure of Yann

The core of the yann toolbox and its operations are built around the yann.network.network class, which is present in the file yann/network.py. Figure A.1 shows the organization of the yann.network .network class. The add_ methods add either a layer or module as nomenclature. The network class can hold many layers and modules in the various connections and architectures that are added using the add_ methods. While prepping the network for learning, we may need only certain modules and layers. The process of preparing the network by selecting and building the training, testing, and validation parts of the network is called *cooking*.

Once cooked, the network is ready for training and testing all by using other methods within the network. The network class also has several properties such as layers, which is a dictionary of the layers that are added to it and params, which is a dictionary of all the parameters. All layers and modules contain a property called id through which they are referred (Figure A.2).

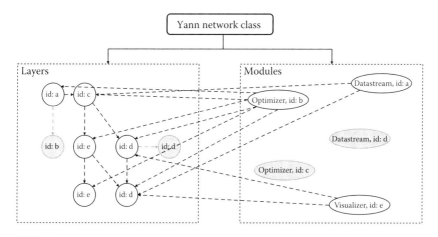

**Figure A.2** A cooked network. The objects that are in gray and shaded are uncooked parts of the network.

**Quick Start with Yann: Logistic Regression**

Let us jump into an example right away and build a logistic regression with yann. The start and the end of the yann toolbox is the yann.network.network class. The yann.network.network object is where all the modeling happens. Start by importing the yann.network.network class and creating a yann.network.network class object:

```
>> from yann.network import network
>> net = network()
```

We have thus created a new network. The network does not have any layers or modules in it. This can be seen by verifying the net.layers property. This will produce an output that is essentially an empty dictionary {}. Let us now add some layers. The toolbox comes with a skdata port to the MNIST dataset of handwritten characters prepackaged for this toolbox. The dataset is typically located at _datasets/_dataset_xxxxx. Refer to the online tutorials on how to convert your own dataset for yann. The first layer that we need to add is an input layer. Every input layer requires a dataset to be associated with it. Let us create this dataset and the layer like so:

```
>> from yann.utils.dataset import cook_mnist
>> cook_mnist()
```

which cooks the dataset. Running this code will print a statement to the following effect:

```
>> Dataset xxxxx is created.
```

The five digits marked xxxxx in the statement are the codeword for the dataset. The actual dataset is located now at _datasets/ _dataset_xxxxx/ from where this code was called. The MNIST dataset is created and stored at this dataset in a format that is configured for yann to work with. Refer to the Tutorials on how to convert your own dataset for yann. The first layer that we need to add is an input layer. Every input layer requires a dataset to be associated with it. Let us create this layer like so:

```
>> dataset_params= { "dataset": "_datasets/_
   dataset_71367", "n_classes" : 10 }
>> net.add_layer(type = "input", dataset_init_args =
   dataset_params)
```

This piece of code creates and adds a new datastream module to the net and wires up the newly added input layer with this datastream. Let us add a logistic regression layer. The default classifier that Yann is setup with is the logistic regression classifier. Refer to the Toolbox documentation or tutorials for other types of layers. Let us create this classifier layer for now:

```
>> net.add_layer(type = "classifier", num_classes = 10)
>> net.add_layer(type = "objective")
```

The layer objective creates the loss function from the classifier layer that can be used as a learning metric. It also provides a scope for other modules such as the optimizer module. Now that our network is created and constructed, we can see that the net objects have layers populated:

```
>> net.layers
>> {'1': <yann.network.layers.classifier_layer object at
   0x7eff9a7d0050>, '0': <yann.network.layers.input_
```

```
layer object at 0x7effa410d6d0>, '2': <yann.network.
layers.objective_layer object at 0x7eff9a71b210>}
```

The keys of the dictionary such as "1," "0," and "2" represent the id of the layer. We could have created a layer with a custom id by supplying an id argument to the add_layer method. Now our network is finally ready to be trained. Before training, we need to build the optimizer and other tools, but for now let us use the default ones. Once all of this is done, before training, yann requires that the network be "cooked":

```
>> net.cook()
```

Depending on the computer, cooking may take a few seconds and might print what it is doing along the way. Once cooked, we may notice for instance that the network has an optimizer module:

```
>> net.optimizer
>> {'main': <yann.network.modules.optimizer object at
   0x7eff9a7c1b10>}
```

To train the model that we have just cooked, we can use the train function that becomes available to us once the network is cooked:

```
>> net.train()
```

This will print the progress for each epoch and will show validation accuracy after each epoch on a validation set that is independent of the training set. By default, the training will run for 40 epochs: 20 on a higher learning rate and 20 more on a fine-tuning learning rate.

Every layer also has an layer.output object. The output can be probed by using the layer.activity method as long as it is directly or indirectly associated with a datastream module through an input layer and the network was cooked. Let us observe the activity of the input layer for a trial. Once trained, we can observe this output. The layer activity will just be a numpy array of numbers, so let us print its shape instead:

```
>> net.layer_activity(id = '0').shape
>> net.layers['0'].output_shape
```

The second line of code will verify the output we produced in the first line. An interesting layer output is the output of the `objective` layer, which will give us the current negative log-likelihood of the network, the one that we are trying to minimize:

```
net.layer_activity(id = '2')
>> array(2.1561384201049805,dtype=float32)
```

Once we are done training, we can run the network feed-forward on the testing set to produce a generalization performance result:

```
>> net.test()
```

We have now successfully used the yann toolbox and implemented logistic regression. Full-fledged code for the logistic regression that we implemented here can be found in `pantry/tutorials/log_reg.py`. That piece of code also has in the commentary other options that could be supplied to some of the function calls we made here that explain the processes better.

## Multilayer Neural Networks

Extending a logistic regression to an MLNN using yann is a simple addition of some hidden layers. Let us add a couple of them. Instead of connecting an input to a classifier as we saw in the regression example, let us add a couple of fully connected hidden layers. Hidden layers can be created using layer type = `dot_product`:

```
>> net.add_layer (type = "dot_product",
            origin ="input",
            id = "dot_product_1",
            num_neurons = 800,
            activation ='relu')

>> net.add_layer (type = "dot_product",
            origin ="dot_product_1",
            id = "dot_product_2",
            num_neurons = 800,
            activation ='relu')
```

Notice the parameters passed. num_neurons is the number of nodes in the layer. Notice also how we modularized the layers by using the id parameter. origin represents which layer will be the input to the new layer. By default, yann assumes all layers are input serially and chooses the last added layer to be the input. Using origin, one can create various types of architectures. In fact, any directed acyclic graphs (DAGs) that could be hand drawn could be implemented. Let us now add a classifier and an objective layer to this:

```
>> net.add_layer (type = "classifier",
                  id = "softmax",
                  origin = "dot_product_2",
                  num_classes = 10,
                  activation = 'softmax',
                  )

>> net.add_layer (type = "objective",
                  id = "nll",
                  origin = "softmax",
                  )
```

Let us create our own optimizer module this time instead of using the yann default. For any module in yann, the initialization can be done using the add_module method. The add_module method typically takes an input type that in this case is optimizer and a set of initialization parameters that in our case is params=optimizer_params. Any module parameters, which in this case is optimizer_params is a dictionary of relevant options. A typical optimizer setup is:

```
>> optimizer_params = {
           "momentum_type"        : 'polyak',
           "momentum_params"      : (0.9,0.95,30),
           "regularization"       : (0.0001,0.0002),
           "optimizer_type"       : 'rmsprop',
           "id"                   : 'polyak-rms'
           }

>> net.add_module (type = 'optimizer',params =
   optimizer_params)
```

We have now successfully added Polyak momentum with RmsProp backpropagation with some $L_1$ and $L_2$ norms. Once the optimizer is added, we can cook and learn:

```
>> learning_rates = (0.05,0.01,0.001)
>> net.cook( optimizer = 'main',
            objective_layer = 'nll',
            datastream = 'mnist',
            classifier = 'softmax',
            )

>> net.train(epochs = (20, 20),
            validate_after_epochs = 2,
            training_accuracy = True,
            learning_rates = learning_rates,
            show_progress = True,
            early_terminate = True)
```

Once done, let us run net.test(). Some new arguments are introduced here and they are for the most part easy to understand in context. epoch represents a tuple that is the number of epochs of training and number of epochs of fine-tuning after that. There could be several of these stages of fine-tuning. Yann uses the term era to represent each set of epochs running with one learning rate. learning_rates indicates the learning rates. The first element of this learning rate is an annealing parameter. learning_rates naturally has a length that is one greater than the number of epochs. show_progress will print a progress bar for each epoch. validate_after_epochs will perform validation after such many epochs on a different validation dataset. The full code for this tutorial with additional commentary can be found in the file pantry.tutorials.mlp.py. Run the code as follows:

```
>> from pantry.examples.mlp import mlp
>> mlp(dataset = 'some dataset created')
```

### Convolutional Neural Network

Now that we are through with the basics, extending this to a LeNet-5 type network is not that difficult. All we need to do is add a few convpool layers, and they can be added using the same add_layer method:

```
>> net.add_layer (type = "conv_pool",
                   origin = "input",
                   id = "conv_pool_1",
                   num_neurons = 20,
                   filter_size = (5,5),
                   pool_size = (2,2),
                    activation = 'tanh',
                    verbose = verbose
                    )

>> net.add_layer ( type = "conv_pool",
                    origin = "conv_pool_1",
                    id = "conv_pool_2",
                    num_neurons = 50,
                    filter_size = (3,3),
                    pool_size = (2,2),
                    activation = 'tanh',
                    verbose = verbose
                    )
```

conv _ pool _ 2 could now be added to the MLNN architecture discussed above and we have a CNN. This CNN would produce a generalization accuracy well over 99% on the MNIST dataset.

The toolbox has several more options that are not discussed here and the descriptions for using them can be found in the yann online documentation. Again code for this is presented in the tutorial's directory.

*Autoencoder*

An autoencoder is a little trickier than the other networks but not too tricky. The initial steps are the same until the first layer (or several layers) of the autoencoder. The decoder layer now needs to take as input the same set of parameters of the encoder layer with a transpose in it. The following code block creates such a layer using yann:

```
>> net.add_layer ( type = "dot_product",
                    origin = "encoder",
                    id = "decoder",
                    num_neurons = 784,
                    activation = 'tanh',
                    input_params = [net.
```

```
        layers['encoder'].w.T,None],
    # Use the same weights but
      transposed for decoder.
    learnable = False,
    verbose = verbose
    )
```

Thus we have added a decoder layer, which will decode the image into 784 outputs and takes as input the transposed weights of the input. The None corresponds to the bias. If needed, we can force the encoder bias on there as well.

To create a loss function, we need to reshape or unflatten the image back into its square shape; we can use the unflatten layer to do this job using the following code:

```
>> net.add_layer ( type = "unflatten",
                   origin = "decoder",
                   id = "unflatten",
                   shape = (28,28,1),
                   verbose = verbose
                   )
```

Once unflattened, the image needs to be compared to the input and this should be the objective to learn from. We can make use of the merge layer and layer _ type arguments in the objective layer to accomplish this task:

```
>> net.add_layer ( type = "merge",
                   origin = ("input","unflatten"),
                   id = "merge",
                   layer_type = "error",
                   error = "rmse",
                   verbose = verbose)

>> net.add_layer ( type = "objective",
                   origin = "merge",
                   id = "obj",
                   objective = None,
                   layer_type = 'generator',
                   verbose = verbose
                   )
```

The autoencoder is thus created. After cooking and training, if we use a visualizer, we can visualize the images being saved and produced.

## Summary

The above examples are intended to quickly demonstrate some of the networks and their types described here. The complete code for all of these is available in the tutorials directory in the code pantry. The toolbox is capable of much more and has more options than those that are described here; however, only selected items have been presented here. Refer to the code documentation at http://www.yann .network for more details and other bleeding-edge developmental tutorials and code.

## References

Abadi, Martin, Agarwal, Ashish, Barham, Paul et al. 2016. *Tensorflow: Large-scale machine learning on heterogeneous distributed systems*. arXiv preprint arXiv:1603.04467.

Bergstra, James, Breuleux, Olivier, Bastien, Frederic et al. 2010. Theano: A CPU and GPU math compiler in Python. *Proceedings of the 9th Python in Science Conference*, pp. 1–7.

Chollet, François. 2015. *Keras*. GitHub repository. https://keras.io/.

Jia, Yangqing, Shelhamer, Evan, Donahue, Jeff et al. 2014. Caffe: Convolutional architecture for fast feature embedding. *Proceedings of the 22nd ACM International Conference on Multimedia*. ACM, pp. 675–678.

Lasagne. n.d. *Lasagne*. http://lasagne.readthedocs.org/en/latest/.

Vedaldi, Andrea and Lenc, Karel. 2015. MatConvNet: Convolutional neural networks for MATLAB. *Proceedings of the 23rd ACM International Conference on Multimedia*. ACM, pp. 689–692.

# POSTSCRIPT

There are several epistemological philosophies on machine learning. In this book, we studied the philosophy of connectionism and its application toward computer vision. We began in Chapter 1 with an introduction to image representations in its various forms and some nonconnectionist image modeling. In Chapter 2, we studied linear regression, optimization, and regularization. Chapter 3 introduced the perceptron or computational neuron, multilayer perceptrons or multilayer neural networks and their learning through the back-propagation algorithm, along with some tricks to make the learning better, faster, and more stable. Chapter 4 introduced the convpool layer and the convolutional neural network (CNN) and studied some of its popular cases. Chapter 5 mainly demonstrated some modern, novel uses of some pretrained networks, which have been among the most popular ways of deploying CNNs in practice. Generative models based on CNNs were also introduced in Chapter 5.

The artificial neural network is one of the classic areas of machine learning and its popularity and usage have gone up and down in past decades. It has often been advertised and sensationalized for what it is not and may have also been overlooked for what it was capable of. It has been largely misunderstood by those who have not diligently studied it. Through its ups and downs, there have been several arguments toward using end-to-end neural networks for computer vision. Among the strongest may be the recent view that "It's learning all the way down" (LeCun, 2015). For those who are interested in a well-written history of connectionism, the authors recommend the book *Talking Nets: An Oral History of Neural Networks* (Anderson and Rosenfeld, 2000).

There have always been philosophical conundrums over connectionism and its implications on the understanding of higher human cognition. A good, albeit controversial source for some of these questions is the article by Fodor and Pylyshyn (1988). Although this article has some well-argued criticisms including the paper by Chalmers (1990) and some of its central premises have been proven wrong, it indeed gives good insight for those who seek some philosophical underpinnings into neural networks. In this book, although we have purposefully stayed away from such philosophies and focused primarily on the engineering of neural networks, the authors would like to point out a modern-day neural network–related reincarnation of a classic debate in philosophy.

Chapter 5 introduced the ideas of fine-tuning a pretrained network versus initializing a new network from scratch. It also motivated several arguments for and against blindly using pretrained networks for computer vision tasks. Most often, it is only through implementation experiences (what is considered "dirty work" [Ng, 2016]) that an engineer could develop the intuition as to when to use pretrained networks and when not to. In the debate over using pretrained networks against training a network from scratch for every task, for most computer vision tasks, some levels of the former seem to be winning this debate as was demonstrated throughout the first part of Chapter 5. Given this, an engineer is always well-advised to try fine-tuning a network pretrained on a related task before attempting to build and train his or her own architecture totally from scratch.

It also appears that, in the debate over the use of neural networks for studying human cognition, the one question that is often raised is the capability of neural networks to generalize the data that is not seen. This is in opposition to the *systematicity* debate whose proponents argue that neural networks merely make associations and do not learn structure or syntax, again from the work of Fodor and Pylyshn (1988). Several early arguments were made in support of neural networks (Niklasson and Gelder, 1994). Neural networks have since demonstrated strong generalization capabilities, the best of which is demonstrated through the use of pretrained networks and through the generator networks generating samples from very complex distributions as in generative adversarial networks (GANs) (Goodfellow et al., 2014). In this widely changing scenario, we require novel ways to think about generalization. Most engineers who develop neural network–related products for certain domains often do not have the amount of data required for training large networks sufficiently in that domain. While pretrained networks are one way out, another practical method is to acquire training data from related domains during training. If training data from a related domain can be obtained easily for learning good representations, which will then be adapted to the testing data's domain, it may provide another way around the problem. Studies on such domain adaptation and transfer learning ideas have become an exciting area to consider.

Another exciting area of research and development in neural networks is reinforcement learning. Reinforcement learning has become popular since demonstrations by companies such as DeepMind and others show that neural networks can learn to play games and can even better the best of humans in such games. Learning to play games with neural networks is not a new concept and has been an active area since its inception. From Arthur Samuel's checkers to Google DeepMind's AlphaGo, neural networks have been shown to be successful in several games. Data are abundant; in the case of reinforcement learning, we can simulate a world that is faster than real time. The community is looking forward to a system that will learn from these domains while ensuring that what was learned here could transfer to the real world. Learning from synthetic data has also sparked the use of GANs to create pretrained networks. Synthesizing data could lead to learning with very small amounts of actual real data and this is currently being actively studied. Another active and possibly the next frontier of neural network research is unsupervised learning. With the rise of mobile phones and image-sharing social media, the amount of image content that is generated has exploded exponentially. Labeling such a vast treasure trove of raw data is a seemingly monumental task and therefore researchers are looking toward unsupervised learning to tap the potential in this data. This would be another exciting area to watch for an interested person.

While not strictly related to neural networks in the technical sense, the recent explosion of interest in neural networks is due to several factors besides neural networks themselves. One that we would like to mention and appreciate is the community's sense of oneness in sharing knowledge and material. The culture of open source software from academics and industries alike has significantly improved the flexibility and speed of discourse and the ease of reproducing some of the results. This culture coupled with tools such as *GitHub*, *travis-ci*, and *readthedocs* for reliable code sharing has had a positive impact and has largely contributed to better collaboration and easy dissemination of scholarly work. We therefore encourage the readers to follow and continue this culture.

While neural networks are proving reliable in the end-to-end learning of several tasks, some tasks still need hand-carved representations or logic. Not all systems could be learned end-to-end purely from the data. While "deep learning can do anything that humans can do in less than a second" (Ng, 2016), this also seems to imply that if a task involves too much cognitive and logical thinking that takes humans a long time to complete, it may not be an easy task for the current deep learning approaches at which to excel. Although informal, we believe that this may still be a good benchmark for the state and accomplishment for current deep learning systems. While we may not be able to prophesize what deep learning and neural networks hold for the future, being able to do anything a human can do in less than a second is clearly not a bad place to begin.

### References

Anderson, James A and Rosenfeld, Edward. 2000. *Talking Nets: An Oral History of Neural networks*. Cambridge, MA: MIT Press.

Chalmers, David. 1990. Why Fodor and Pylyshyn were wrong: The simplest refutation. *Proceedings of the Twelfth Annual Conference of the Cognitive Science Society,* pp. 340–347.

Fodor, Jerry A and Pylyshyn, Zenon W. 1988. Connectionism and cognitive architecture: A critical analysis. *Cognition* (Elsevier) 28: 3–71.

Goodfellow, Ian, Pouget-Abadie, Jean, Mirza, Mehdi et al. 2014. Generative adversarial nets. In *Advances in Neural Information Processing Systems,* pp. 2672–2680.

LeCun, Yann. 2015. It's learning all the way down. *International Conference on Computer Vision.*

Ng, Andrew. 2016. Nuts and bolts of deep learning. *NIPS 2016, Deep/Bay Area Deep Learning School.*

Niklasson, Lars F and Gelder, Tim. 1994. On being systematically connectionist. *Mind & Language* 9: 288–302.

# Index